The ghost in the atom

A discussion of the mysteries of quantum physics

The ghost in the atom

P. C. W. DAVIES
*Professor in the Department of Physics and Mathematical Physics,
University of Adelaide*

J. R. BROWN
Radio Producer in the BBC Science Unit, London

CAMBRIDGE
UNIVERSITY PRESS

Published by the Press Syndicate of the University of Cambridge
The Pitt Building, Trumpington Street, Cambridge CB2 1RP
40 West 20th Street, New York, NY 10011–4211, USA
10 Stamford Road, Oakleigh, Melbourne 3166, Australia

First published 1986
Reprinted 1987, 1988, 1989, 1991
Canto edition 1993

Printed in Great Britain at the University Press, Cambridge

British Library cataloguing in publication data

The Ghost in the atom: a discussion of the
mysteries of quantum physics.

1. Quantum theory
I. Davies, P. C. W.   II. Brown, J.
530.1'2   QC174.12

Library of Congress cataloguing in publication data

Davies, P. C. W.
The ghost in the atom.

Bibliography
Includes index.
1. Quantum theory.   2. Physicist—Interviews.
I. Brown, J. (Julian), 1957– .   II. Title.
QC174.12D365 1986   530.1'2   85–25478

ISBN 0 521 30790 2 hardback
ISBN 0 521 45728 9 paperback

PN

CONTENTS

FOREWORD

Niels Bohr once remarked that anybody who is not shocked by quantum theory has not understood it. Certainly a powerful sense of shock and bewilderment reverberated among his contemporaries in the 1920s when the full implications of the theory began to emerge. Not only did quantum theory fly in the face of classical nineteenth-century physics but it also radically transformed scientists' outlook on our relationship with the material world. For, according to Bohr's interpretation of the theory, the existence of the world 'out there' is not something that enjoys an independence of its own, but is inextricably tied up with our perceptions of it.

Not surprisingly, some physicists found such an idea hard to swallow. Ironically, having played a significant part in the early development of quantum theory, Albert Einstein became its foremost critic. Until his death in 1955, he was convinced that an essential ingredient was missing from the formulation of quantum theory; without this ingredient he argued, our description of matter on the atomic scale would inevitably remain intrinsically uncertain and therefore incomplete. In the course of a long friendship with Bohr, Einstein repeatedly tried to demonstrate the incompleteness of quantum theory. He produced a number of highly ingenious arguments, some of which caused considerable concern among scientists. But each time Bohr quickly managed to find an elegant and persuasive refutation. Gradually, the feeling grew that Einstein's quest to exorcise the ghost in the atom had been in vain.

But today the quantum controversy is far from over. In recent

years a series of experimental tests has been carried out, culminating in those of Alain Aspect and his colleagues in France –
tests which promised to cast new light on the Bohr–Einstein
debate.

The resurgence of interest in the interpretation of quantum
theory prompted me (J.B.) to consider making a radio documentary on the subject. I discussed the idea with Professor Paul
Davies, who agreed to present a programme for BBC Radio 3. We
interviewed several leading physicists who have taken a particular interest in the conceptual foundations of quantum mechanics,
to find out what they made of Aspect's results and other recent
developments in quantum theory.

Owing to the naturally quite limited time available within a
documentary format, only brief segments of the interviews could
be used in the final programme. Nevertheless, the Radio 3
broadcast of the 'The Ghost in the Atom' provoked a great deal of
interest and subsequently we felt that it would be well worth
while publishing the interviews in a fuller and more permanent
form.

With the exception of Chapter 1, the contents of this book are
based upon the transcripts of the original radio interviews. In
editing them, we have been obliged to make some amendments
to render the dialogues more suitable for the printed page, but
we have endeavoured to do this without sacrificing too much of
their conversational character. This book is intended for the
general reader, and we have therefore written Chapter 1 as an
introduction to the ideas discussed within the interviews. If you
are already familiar with many of these, you may wish to jump
directly to Chapter 2 and refer to the index or glossary for
explanations of any technical terms or arguments.

A final thought and a note of caution; when we commissioned
the interviews, several of our contributors (who shall remain
nameless!) expressed the view that there is now no real doubt
over how quantum theory should be interpreted. At the very
least, we hope this book will show that there is little justification
for such complacency.

We are greatly indebted to all our contributors and especially Sir Rudolf Peierls for his critical reading of Chapter 1. We would also like to thank Mandy Eustace for performing the difficult task of transcribing the contents of the original audio tapes.

January 1986
J. Brown
P. C. W. Davies

1

The strange world of the quantum

What is quantum theory?

The word 'quantum' means 'a quantity' or 'a discrete amount'. On an everday scale we are accustomed to the idea that the properties of an object such as its size, weight, colour, temperature, surface area, and motion are all qualities which can vary from one object to another in a smooth and continuous way. Apples, for example, come in all manner of shapes, sizes and colours without any noticeable gradations in between.

On the atomic scale, however, things are very different. The properties of atomic particles such as their motion, energy and spin do not always exhibit similar smooth variations, but may instead differ in discrete amounts. One of the assumptions of classical Newtonian mechanics was that the properties of matter are continuously variable. When physicists discovered that this notion breaks down on the atomic scale they had to devise an entirely new system of mechanics – quantum mechanics – to take account of the lumpiness which characterizes the atomic behaviour of matter. Quantum theory, then, is the underlying theory from which quantum mechanics is derived.

Considering the success of classical mechanics in describing the dynamics of everything from billiard balls to stars and planets, it is not surprising that its replacement by a new system of mechanics on the atomic scale was considered to be a revolutionary departure. Nevertheless, physicists rapidly proved the value of the theory by explaining a wide range of otherwise incomprehensible phenomena, so much so that today quantum theory is often cited as the most successful scientific theory ever produced.

Origins

Quantum theory had its first faltering beginnings in the year 1900, with the publication of a paper by the German physicist Max Planck. Planck addressed himself to what was still an unsolved problem of nineteenth-century physics, concerning the distribution of radiant heat energy from a hot body among various wavelengths. Under certain ideal conditions the energy is distributed in a characteristic way, which Planck showed could only be explained by supposing that the electromagnetic radiation was emitted from the body in discrete packets or bundles, which he called quanta. The reason for this jerky behaviour was unknown, and simply had to be accepted *ad hoc*.

In 1905 the quantum hypothesis was bolstered by Einstein, who successfully explained the so-called photoelectric effect in which light energy is observed to displace electrons from the surfaces of metals. To account for the particular way this happens, Einstein was compelled to regard the beam of light as a hail of discrete particles later called photons. This description of light seemed utterly at odds with the traditional view, in which light (in common with all electromagnetic radiation) consists of continuous waves which propagate in accordance with Maxwell's celebrated electromagnetic theory, firmly established half a century before. Indeed, the wave nature of light had been demonstrated experimentally as long ago as 1801 by Thomas Young using his famous 'two-slit' apparatus.

The wave–particle dichotomy, however, was not restricted to light. Physicists were at that time also concerned about the structure of atoms. In particular, they were puzzled by how electrons could go round and round a nucleus without emitting radiation, since it was known from Maxwell's electromagnetic theory that when charged particles move along curved paths they radiate electromagnetic energy. If this were to occur continuously, the orbiting atomic electrons would rapidly lose energy and spiral into the nucleus (see Fig. 1).

In 1913 Niels Bohr proposed that atomic electrons are also 'quantized', in that they can reside without loss of energy in certain fixed energy levels. When electrons jump between the levels, electromagnetic energy is released or absorbed in discrete quantities. These packets of energy are, in fact, photons.

The reason why the atomic electrons should behave in this discontinuous fashion was not revealed, however, until somewhat later, when the wave nature of matter was discovered. The experimental work of Clinton Davisson and others and the theoretical work of Louis de Broglie led to the idea that electrons as well as photons can behave both as waves and as particles, depending on the particular circumstances. According to the wave picture, the atomic energy levels Bohr proposed correspond to stationary or standing wave patterns around the nucleus. Much as a cavity can be made to resonate at different discrete musical notes, so the electron waves vibrate with certain well-defined energy patterns. Only when the patterns shift, corresponding to a transition from one energy level to another, does an electromagnetic disturbance ensue, with radiation being emitted or absorbed.

Fig. 1. Collapse of the classical atom. (a) The theories of Newton and Maxwell predict that an orbiting atomic electron will steadily radiate electromagnetic waves, thereby losing energy and spiralling into the nucleus. (b) The quantum theory predicts the existence of discrete non-radiating energy levels in which the wave associated with the electron just 'fits' around the nucleus, forming standing wave patterns reminiscent of the notes on a musical instrument. (The wave must 'fit' in the radial direction too.)

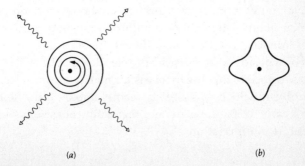

(a)   (b)

It soon became apparent that not only electrons but all sub-atomic particles are subject to similar wavelike behaviour. Evidently the traditional laws of mechanics as formulated by Newton, as well as Maxwell's laws of electromagnetism, fail completely in the microworld of atoms and subatomic particles. By the mid-1920s, a new system of mechanics – quantum mechanics – had been developed independently by Erwin Schrödinger and Werner Heisenberg to take account of this wave–particle duality.

The new theory was spectacularly successful. It rapidly helped scientists to explain the structure of atoms, radioactivity, chemical bonding and the details of atomic spectra (including the effects of electric and magnetic fields). Further elaborations of the theory by Paul Dirac, Enrico Fermi, Max Born and others eventually led to satisfactory explanations of nuclear structure and reactions, the electrical and thermal properties of solids, superconductivity, the creation and annihilation of elementary particles of matter, the prediction of the existence of antimatter, the stability of certain collapsed stars and much else. Quantum mechanics also made possible major developments in practical hardware, including the electron microscope, the laser and the transistor. Exceedingly delicate atomic experiments have confirmed the existence of subtle quantum effects to an astonishing degree of accuracy. No known experiment has contradicted the predictions of quantum mechanics in the last 50 years.

This catalogue of triumphs singles out quantum mechanics as a truly remarkable theory – a theory that correctly describes the world to a level of precision and detail unprecedented in science. Nowadays, the vast majority of professional physicists employ quantum mechanics, if not almost unthinkingly, then with complete confidence. Yet this magnificent theoretical edifice is founded on a profound and disturbing paradox that has led some physicists to declare that the theory is ultimately meaningless.

The problem, which was already readily apparent in the late 1920s and early 1930s, concerns not the technical aspects of the theory but its interpretation.

Waves or particles?

The peculiarity of the quantum is readily apparent from the way that an object such as a photon can display both wave-like and particle-like properties. Photons can be made to produce diffraction and interference patterns, a sure test of their wave-like nature. On the other hand, in the photoelectric effect, photons knock electrons out of metals after the fashion of a coconut-shy. Here, the particle model of light seems to be more appropriate.

The co-existence of wave and particle properties leads quickly to some surprising conclusions about nature. Let us take a familiar example. Suppose that a beam of polarized light encounters a piece of polarizing material (see Fig. 2). Standard electromagnetic theory predicts that if the plane of polarization of the light is parallel to that of the material, all the light is transmitted. On the other hand, if the angles are perpendicular, no light is transmitted. At intermediate angles some light is transmitted; for example, at 45° the transmitted light has precisely half the intensity of the original beam. Experiment confirms this.

Fig. 2. Breakdown of predictability. (a) Classically, the polarized light wave will pass through the polarizer with a reduced intensity $cos^2 \theta$, emerging polarized in the 'vertical' direction. Viewed as a flux of identical photons, this phenomenon can be explained only by supposing that some photons are passed and others blocked, unpredictably, with probabilities $cos^2 \theta$ and $sin^2 \theta$, respectively. (b) Note that the incident wave could be regarded as a superposition of 'vertically' and 'horizontally' polarized waves.

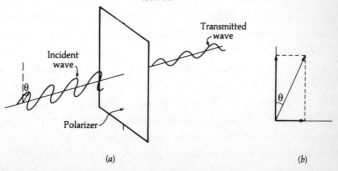

Incident wave

Transmitted wave

Polarizer

θ

(a)

(b)

Now, if the intensity of the incoming beam is reduced so that only one photon at a time passes through the polarizer, we are faced with a puzzle. Because a photon cannot be divided up, any given photon must be either passed or blocked. With the angle set at 45°, on average half the photons must get through, while the other half are blocked. But which photons get through and which do not? As all photons of the same energy are supposed to be identical and hence indistinguishable, we are forced to conclude that the passage of photons is a purely random process. Although any given photon has a 50–50 chance (a probability of $\frac{1}{2}$) of getting through, it is impossible to predict in advance which particular ones will do so. Only the betting odds can be given. As the angle is varied so the probability can range from zero to one.

The conclusion is intriguing and yet disconcerting. Before the discovery of quantum physics the world was thought to be completely predictable, at least in principle. In particular, if identical experiments were performed, identical results were expected. But, in the case of the photons and the polarizer, one might very well find that two identical experiments produced different results, as one photon passed through the polarizer while another identical photon was blocked. Evidently the world is not wholly predictable after all. Generally we cannot know until after an observation has been made what the fate of a given photon will be.

These ideas imply that there is an element of uncertainty in the microworld of photons, electrons, atoms, and other particles. In 1927 Heisenberg quantified this uncertainty in his famous uncertainty principle. One expression of the principle concerns attempts to measure the position and motion of a quantum object simultaneously. Specifically, if we try to locate an electron, say, very precisely, we are forced to forgo information about its momentum. Conversely, we can measure the electron's momentum accurately, but then its position becomes indeterminate. The very act of trying to pin down an electron to a specific place introduces an uncontrollable and indeterminate disturbance to its motion, and vice versa. Furthermore, this inescapable constraint on our knowledge of the electron's motion and location is not merely the result of experimental clumsiness; it is inherent in nature. Apparently the electron simply *does not possess* both a position and a momentum simultaneously.

It follows that there is an intrinsic fuzziness in the microworld that is manifested whenever we attempt to measure two incompatible observable quantities, such as position and momentum. Among other things, this fuzziness demolishes the intuitive idea of an electron (or photon, or whatever) moving along a distinct path or trajectory in space. For a particle to follow a well-defined path, at each instant it must possess a location (a point on the path) and a motion (tangent vector to the path). But a quantum particle cannot have both at once.

In daily life we take it for granted that strict laws of cause and effect direct the bullet to its target or the planet in its orbit along a precisely defined geometrical path in space. We would not doubt that when the bullet arrives at the target its point of arrival represents the end-point of a continuous curve which started at the barrel of the gun. Not so for electrons. We can discern a point of departure and a point of arrival, but we cannot always infer that there was a definite route connecting them.

Seldom is this fuzziness more apparent than in the famous

Fig. 3. *Waves or particles? In this two-slit experiment electrons or photons from the source pass through two nearby apertures in screen A and travel on to strike screen B, where their rate of arrival is monitored. The observed pattern of varying intensity indicates a wave interference phenomenon.*

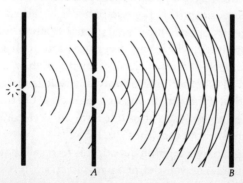

two-slit experiment of Thomas Young (see Fig. 3). Here a beam of photons (or electrons) from a small source travels towards a screen punctured by two narrow apertures. The beam creates an image of the holes on a second screen. The image consists of a distinct pattern of bright and dark 'interference fringes', as waves passing through one hole encounter those from the other hole. Where the waves arrive in step, reinforcement occurs; where they are out of step, cancellation occurs. Thus is the wave-like nature of photons or electrons clearly demonstrated.

But the beam can instead be considered as consisting of particles. Suppose the intensity is again reduced so much that only one photon or electron traverses the apparatus at a time. Naturally each arrives at a definite point on the image screen. It can be recorded as a little speck. Other particles arrive elsewhere leaving their own specks. The effect at first seems random. But a pattern begins to build up in a speckled kind of way. Each particle is directed not by an imperative to a particular place on the image screen but by the 'law of averages'. When a large number of particles has traversed the system, an organized pattern is created. This is the interference pattern. Thus, any given photon or electron does not make a pattern; it makes only a single spot. Yet each electron or photon, while apparently free to go anywhere, cooperates in such a way as to build up the pattern in a probabilistic fashion.

Now, if one of the two apertures is blocked, the average behaviour of the electrons or photons changes dramatically; indeed, the interference pattern disappears. Nor can it be reconstructed by superimposing the two patterns obtained by recording the images from each individual slit acting alone. Interference only presents itself when both apertures are open simultaneously. Hence, each photon or electron must somehow *individually* take account of whether both or only one hole is open. But how can they do this if they are indivisible particles? On the face of it, each particle can only go through one slit. Yet somehow the particle 'knows' about the other slit. How?

One way of answering this question is to recall that quantum particles do not have well-defined paths in space. It is sometimes convenient to think of each particle as somehow possessing an infinity of different paths, each of which contributes to its behaviour. These paths, or routes, thread through both holes in the screen, and encode information about each. This is how the particle can keep track of what is happening throughout an extended region of space. The fuzziness in its activity enables it to 'feel out' many different routes.

Suppose a disbelieving physicist were to station detectors in front of the two holes to ascertain in advance towards which hole a particular electron was heading. Could not the physicist then suddenly block the other hole without the electron 'knowing', leaving its motion unaltered? If we analyse the situation, taking into account Heisenberg's uncertainty principle, then we can see that nature outmanoeuvres the wily physicist. In order for the position of each electron to be measured accurately enough to discern the hole it is approaching, the electron's motion is so disturbed that the interference pattern defiantly vanishes! The very act of investigating where the electron is going ensures that the two-hole cooperation fails. Only if we decide not to trace the electron's route will its 'knowledge' of both routes be displayed.

A further intriguing consequence of the above dichotomy has been pointed out by John Wheeler. The decision either to perform the experiment to determine the electron's route, or to relinquish this knowledge and experiment instead with an interference pattern, can be left until *after* any given electron has already traversed the apparatus! In this so-called 'delayed-choice' experiment, it appears that what the experimenter decides now can in some sense influence how quantum particles shall have behaved in the past, though it must be emphasized that the inherent unpredictability of all quantum processes forbids this arrangement from being used to send signals backwards in time or to in any way 'alter' the past.

An idealized arrangement designed to carry out a related delayed-choice experiment (with photons rather than electrons) is shown in Fig. 4, and forms the basis of a practical experiment performed recently by Caroll Alley and his colleagues at the University of Maryland. Laser light incident on a half-silvered mirror *A* divides into two beams analogous to the two paths through the slits in Young's experiment. Further reflections at mirrors *M* redirect the beams so that they cross and enter photon detectors 1 and 2, respectively. In this arrangement a detection of a given photon by either 1 or 2 suffices to determine which of the two alternative routes the photon will have travelled.

If, now, a second half-silvered mirror *B* is inserted at the crossing point (see Fig. 4) the two beams are recombined, part along the route into 1 and part along the route into 2. This will cause wave interference effects, and the strengths of the beams going into 1 and 2 respectively will then depend on the relative phases of the two beams at the point of recombination. These phases can be altered by adjusting the path lengths, thereby essentially scanning the interference pattern. In particular it is possible to arrange the phases so that destructive interference leads to zero beam strength going into 1, with 100% of the light going into 2. With this arrangement the system is analogous to the original Young experiment, for which it is not possible to specify which of the two routes has been taken by any given photon. (Loosely speaking, each photon takes both routes.)

Fig. 4. *Schematic diagram showing the layout of a practical version of Wheeler's delayed-choice experiment.*

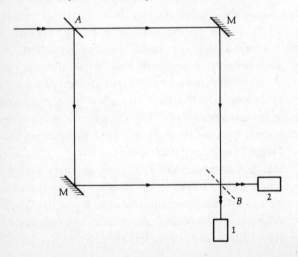

Now the crucial point is that the decision of whether or not to insert the second half-silvered mirror *B* can be left until a given photon has almost arrived at the cross-over point. In other words, whether the photon *shall* have traversed the system either by one route or 'both routes' is determined only *after* the traverse has taken place.

What does it all mean?

The fact that electrons, photons and other quantum objects behave sometimes like waves and sometimes like particles often prompts the question of what they 'really' are. The conventional position regarding questions of this sort draws upon the later work of Bohr, who believed he had discovered a consistent interpretation of quantum mechanics. This is usually referred to as the Copenhagen interpretation, so named after Bohr's physics institute in Denmark, which he founded in the 1920s.

According to Bohr, it is meaningless to ask what an electron 'really' is. Or at least, if you ask the question, physics cannot supply the answer. Physics, he declared, tells us not about what *is*, but what we can *say* to each other concerning the world. Specifically, if a physicist carries out an experiment on a quantum system, provided a full specification of the experimental set-up is given, physics can then make a meaningful prediction about what he may observe, and thence communicate to his fellows in a well-understood language.

In Young's experiment, for example, we have a clear choice. Either we can leave the electrons or photons alone, and observe an interference pattern. Or we can take a peek at the particles' trajectories and wash out the pattern. The two situations are not contradictory, but complementary.

Similarly there is a position–momentum complementarity. We can choose to measure the position of a particle in which case its momentum is uncertain, or we can measure the momentum and trade-off knowledge of its position. Each quality – position, momentum – constitutes a complementary aspect of the quantum object.

Bohr elevated these ideas to a principle of *complementarity*. In wave–particle duality, for example, the wave and particle properties of a quantum object constitute complementary aspects of its behaviour. He argued that we should never encounter any experiments in which these two distinct behaviours conflict with each other.

A profound consequence of Bohr's ideas is that the traditional Western concept of the relationship between macro and micro, the whole and its parts, is radically altered. Bohr claimed that before you can make sense of what an electron is doing you have to specify the total experimental context; say what you are going to measure, how your apparatus is organized, and so on. So the quantum reality of the microworld is inextricably entangled with the organization of the macroworld. In other words, the part has no meaning except in relation to the whole. This holistic character of quantum physics has found considerable favour among followers of Eastern mysticism, the philosophy embodied in such oriental religions as Hinduism, Buddhism and Taoism. Indeed, in the early days of quantum theory many physicists, including Schrödinger, were quick to draw parallels between the quantum concept of part and whole, and the traditional oriental concept of the unity and harmony of nature.

Central to Bohr's philosophy is the assumption that uncertainty and fuzziness are intrinsic to the quantum world and not merely the result of our incomplete perception of it. This is quite a subtle matter. We know of many systems which are unpredictable: the vicissitudes of the weather, stockmarkets and roulette wheels, for example, are familiar enough. Yet these do not force us to make a radical reappraisal of the laws of physics. The reason is that the unpredictability of most things in everyday life can be traced to the fact that we do not have enough information to compute their behaviour at the level of detail necessary for an accurate prediction. In the case of roulette, say, we resort to a statistical description. Likewise in classical thermodynamics, the collective behaviour of myriads of molecules can successfully be described in an average way using statistical mechanics. However, the fluctuations about computed mean values are not intrinsically indeterminate in that case because, in principle, the complete mechanical description of every participating molecule could be given (ignoring for this example quantum effects!).

When the information concerning some dynamical variables is discarded, an element of vagueness and uncertainty is introduced into our description of the system. However, we know that this fuzziness is really the result of the activity of all those variables we have chosen to ignore. We might call them 'hidden variables'. They are always there, but our observations may be too crude to reveal them. Thus the measurement of gas pressure is too coarse to reveal individual molecular motions.

Why can we not attribute quantum fuzziness to a deeper level of hidden variables? Such a theory would enable us to picture the chaotic, apparently indeterminate cavorting of quantum particles as driven by a substratum of completely deterministic forces. The fact that we seem to be unable to determine both the position and momentum of an electron simultaneously might then be attributed to the crude nature of our apparatus which is as yet unable to probe the finer level of this substratum.

Einstein was convinced that something like this must be the case; that ultimately a classical world of familiar cause and effect underlies the madhouse of the quantum. He endeavoured to construct thought experiments to test the idea. The most refined of these he presented in a now famous paper written in 1935 with Boris Podolsky and Nathan Rosen.

The Einstein–Podolsky–Rosen (EPR) experiment

The purpose of this thought experiment was to expose the profound pecularities of the quantum description of a physical system extended over a large region of space. The experiment invites us to consider cheating the Heisenberg uncertainty principle by sneaking a look at both the position and momentum of a particle simultaneously. The strategy employed is to use an accomplice particle to perform a measurement by proxy on the particle of interest.

Suppose a single stationary particle explodes into two equal fragments, *A* and *B* (see Fig. 5). Heisenberg's uncertainty principle apparently forbids us from knowing the position and momentum of either *A* or *B* simultaneously. However, because of the law of action and reaction (conservation of momentum), a measurement of *B*'s momentum can be used to deduce *A*'s momentum. Similarly, by symmetry, *A* will have moved a distance equal to that of *B* from the point of explosion, so a measurement of *B*'s position reveals *A*'s position.

An observer at *B* is free, at his whim, to observe either the momentum or the position of *B*. As a result he will know either the momentum or the position of *A*, according to his choice. Thus, a subsequent observation of either *A*'s momentum or *A*'s position will now have a predictable result.

Einstein argued 'If, without in any way disturbing a system, we can predict with certainty . . . the value of a physical quantity, then there exists an element of physical reality corresponding to this physical quantity.' He therefore concluded that, in the situation described, the particle *A* must possess a real momentum or a real position, according to the choice of the observer at *B*.

Now the crucial point is this. If *A* and *B* have flown a very long way apart then one would be reluctant to suppose that a measurement carried out on *B* can affect *A*. At the very least, *A* cannot be directly affected instantaneously, because according to the special theory of relativity no physical signal or influence can travel faster than light; so *A* cannot 'know' that a measurement has been performed on *B* until at least the light travel time between them. In principle this could be billions of years!

Fig. 5. Two equal mass fragments flying apart from a common centre (assumed at rest) have equal and opposite momenta and are always equidistant from the centre. Hence a measurement of either the momentum or position of A *reveals the momentum or position of* B.

A   B

Bohr rejected Einstein's reasoning by reiterating his Copenhagen philosophy, that the microscopic properties of a quantum particle must be viewed against the total macroscopic context. In this case a distant but correlated accomplice article, subjected to measurements, forms an inseparable part of the quantum system. Although no direct signal or influence can travel between A and B, that does not mean, according to Bohr, that you can ignore measurements carried out on B when discussing the circumstances of A. So, although no actual physical force is transmitted between A and B, they seem to *cooperate* in their behaviour in a sort of conspiracy.

Einstein found this idea of two widely separated particles conspiring to give coordinated results of apparently independent measurements performed on each too much to swallow, deriding it as 'ghostly action at a distance'. He wanted his objective reality to be localized on each particle, and it was this locality that was eventually to bring his ideas into conflict with quantum mechanics. What was needed was a practical experimental test that could discriminate between Bohr's and Einstein's views by revealing the cooperation, or ghostly action at a distance, in action. But such a development had to wait half a century.

Bell's theorem

In 1965 John Bell studied the problem of two-particle quantum systems and was able to prove a powerful mathematical theorem which turned out to be of crucial importance in setting up a practical experimental test. The theory is essentially independent of the nature of the particles or the details of the forces that act on them, and focusses instead on the rules of logic that govern all measurement processes. To give a simple example of the latter, a census of the population of Britain cannot possibly find that the number of black people is greater than the number of black men plus the number of women of all races.

Bell investigated the correlations that could exist between the results of measurements carried out simultaneously on two separated particles. These measurements might be on particle positions, momenta, spin, polarization, or other dynamical properties. Many researchers have adopted polarization as a convenient means of studying EPR correlations. Suppose a parent particle with zero angular momentum decays into two photons A and B. By the laws of conservation, one photon must have the same polarization as the other. This can be confirmed by stationing measuring devices perpendicular to the paths of the particles and measuring the polarization in a certain common direction, say 'up'. It is indeed found that, when particle A is passed by its polarizer, B is always passed too. A 100% *correlation* is found. Conversely, if the polarizers are arranged perpendicular to each other, every time A is passed B is blocked. This time there is 100% *anti-correlation*. There is nothing mysterious about this; it would also be true in ordinary classical mechanics.

The crucial test comes when the polarization measuring devices are oriented obliquely to each other (see Fig. 6). We would now expect some result intermediate between complete correlation and complete anti-correlation, depending on the angles chosen. These may be varied both parallel and perpendicular to the line of flight of the particles, and they could be varied at random from one measurement to the next.

Bell set out to discover the theoretical limits on the extent to which the results of such measurements can be correlated. Suppose, for example, that Einstein had been basically correct, and that quantum behaviour is really the product of a substratum of chaotic classical forces. Suppose also that faster-than-light signalling is forbidden in accordance with the rules of relativity theory. Properly formulated, the first assumption is usually what

Fig. 6. Bell's theorem applied to two oppositely directed photons from a common source predicts a limit to the degree of correlation permitted in the results of polarization measurements performed separately on each.

Source

is meant by 'reality', because it affirms that quantum objects *really do* possess *all* dynamical attributes in a well-defined sense at all times. The second assumption is termed 'locality' or sometimes 'separability' because it forbids objects from instantaneously exerting physical influences on each other when they are spatially separated, i.e. not at the same location.

Subject to the double assumption of 'local reality', and further assuming that the conventional rules of logical reasoning do not founder on the rocks of quantum uncertainty, Bell was able to establish a strict limit on the possible level of correlation for simultaneous two-particle measurement results. The whole point of the exercise then is this. Quantum mechanics *à la* Bohr predicts that, under some circumstances the degree of cooperation should *exceed* Bell's limit. That is, the conventional view of quantum mechanics requires a degree of cooperation (or conspiracy) between separated systems in excess of that logically permitted in any 'locally real' theory. Bell's theorem thus opens the way for a direct test of the foundations of quantum mechanics, and the decisive discrimination between Einstein's idea of a locally real world, and Bohr's conception of a somewhat ghostly world full of subatomic conspiracy.

Aspect's experiment

A number of experiments have been conducted in an attempt to test Bell's inequality. The most successful of these was reported by A. Aspect, J. Dalibard and G. Roger in *Physical Review Letters* (vol. **39**, p. 1804) in December 1982.

The experiment consisted of polarization measurements made on pairs of oppositely moving photons emitted simultaneously in single transitions by calcium atoms. The experimental arrangement is shown in Fig. 7.

In the diagram, the source S used a beam of calcium atoms excited by a pair of lasers (i.e. two-photon excitation) to a state (S state) that could only decay again by a two-photon 'cascade'. About 6 m on either side of the source there was located an acousto-optical switching device. The principle employed was to

Fig. 7. Aspect's experimental arrangement. Pairs of photons from the source S travel several metres to the acousto-optical switches. The route of a photon beyond the switch determines which of the differently oriented polarizers it will encounter. The photons are detected using photo-multipliers (PM) and coincidences between the various channels are monitored electronically. The photograph shows the actual experimental set-up (courtesy of A. Aspect).

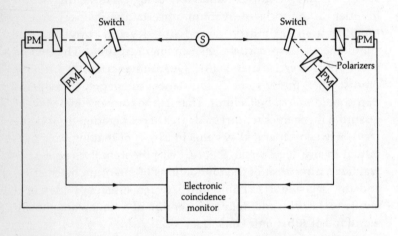

exploit the fact that the refractive index of water will vary slightly with compression.

In the switch, an ultrasonic standing wave at about 25 MHz was established using oppositely directed transducers. By arranging for the photons to encounter the switch at near the critical angle for total internal reflection, it was possible to switch from transmission to reflection conditions at each half-cycle of the sound wave, i.e. at 50 MHz.

The photons, emerging either along the line of the incident path (after transmission) or deflected (by reflection) then encountered polarizers, which would either transmit or block them with certain definite probabilities. These polarizers were oriented at different angles relative to the polarization of the photons. The photons' fate was then monitored by stationing photomultiplier detectors beyond these polarizers. The set-up was identical on both sides of the source.

The experiment was performed by monitoring electronically the fate of each pair of photons and assessing the level of correlation. The unique and essential feature of this experiment is the ability to change at random, while the photons are in mid-flight, the subsequent path of the photons, i.e. to which polarizer they shall be directed. This is equivalent to re-orienting the polarizers on each side of the source so rapidly that no signal could have time enough to pass from one to the other, even at the speed of light.

Each switching event took about 10 ns, which should be compared with the lifetime of the photons' emission (5 ns) and the travel time of the photons (40 ns).

In practice, the switching was not strictly random. The standing waves were generated independently at different frequencies. The difference between this and truly random switching is irrelevant except in the case of the most contrived 'conspiracy' theories of hidden variables.

The authors report that a typical run lasted 12 000 s, divided equally between the arrangement as described above, another in which all the polarizers were removed, and a third in which one polarizer on each side was removed. This enabled the results to be corrected for systematic errors.

uniformly throughout the box. Suppose now that an impenetrable screen is inserted down the middle of the box, dividing it into two chambers. Obviously the electron can only be in *either* one chamber *or* the other. However, unless we look and see which, the *wave* will still be in both chambers. On observation the electron will be revealed to be in one particular chamber. At that very instant (according to the rules of quantum mechanics) the wave abruptly disappears from the empty chamber, even if that chamber has remained isolated throughout! It is as though, prior to the observation, there are two nebulous electron 'ghosts' each inhabiting one chamber waiting for an observation to turn one of them into a 'real' electron, and simultaneously to cause the other to vanish completely.

This example also nicely illustrates the non-locality of quantum mechanics. Suppose the two chambers, *A* and *B*, are disconnected and moved a long way apart (say one light year), then *A* is inspected by an observer and found to contain the particle. Instantaneously the quantum wave in *B* vanishes, even though it is a light year away. (It must be repeated, however, that this arrangement cannot be used to signal faster than light, on account of the unpredictable nature of each observation.)

In general, a quantum system will be in a state consisting of a collection (perhaps infinite in number) of quantum states superimposed. A simple example of such a superposition was given above, involving two disconnected wave patterns, one in each chamber. A more typical example is that in Young's two-slit experiment, where waves from both slits actually overlap and interfere.

We encountered this sort of superposition earlier, in the discussion of polarized light passing through an obliquely oriented polarizer. If the incoming light wave is at 45° to the polarizer, we may consider it to consist of two equal-strength waves combined coherently with polarizations at right angles to each other, as shown in Fig. 2. The wave parallel to the polarizer will be transmitted, the other will be blocked. We could regard a quantum state containing one photon polarized at 45° to the polarizer as a superposition of two 'ghosts' or 'potential' photons, one with parallel polarization enabling it to get through the polarizer, the other with perpendicular polarization, preventing it from getting through. When the measurement is finally made, one of these two 'ghosts' gets promoted to a 'real' photon and the other vanishes. Suppose the measurement shows that the photon passes through the polarizer. The *ghost* photon that is parallel to the polarizer prior to the measurement thus becomes the *real* photon. But we cannot say that this photon 'really existed' *prior* to the measurement. All that can be said is that the system was in a superposition of two quantum states, neither of which possessed privileged status.

The physicist John Wheeler likes to tell a delightful parable which nicely illustrates the peculiar status of a quantum particle prior to measurement. The story concerns a version of the game of 20 questions:

> Then my turn came, fourth to be sent from the room so that Lothar Nordheim's other fifteen after-dinner guests could consult in secret and agree on a difficult word. I was locked out unbelievably long. On finally being readmitted, I found a smile on everyone's face, sign of a joke or a plot. I nevertheless started my attempt to find the word. 'Is it animal?' 'No.' 'Is it mineral?' 'Yes.' 'Is it green?' 'No.' 'Is it white?' 'Yes.' These answers came quickly. Then the questions began to take longer in the answering. It was strange. All I wanted from my friends was a simple yes or no. Yet the one queried would think and think, yes or no, no or yes, before responding. Finally I felt I was getting hot on the trail, that the word might be 'cloud'. I knew I was allowed only one chance at the final word. I ventured it: 'Is it cloud?' 'Yes,' came the reply, and everyone burst out laughing. They explained to me there had been no word in the room. They had agreed not to agree on a word. Each one questioned could answer as he pleased – with the one requirement that he should have a word in mind compatible with his own response and all that had gone before. Otherwise, if I challenged, he lost. The surprise version of the game of twenty questions was therefore as difficult for my colleagues as it was for me.

What is the symbolism of the story? The world, we once believed, exists 'out there' independent of any act of observation. The electron in the atom we once considered to have at each moment a definite position and a definite momentum. I, entering, thought the room contained a definite word. In actuality the word was developed step by step through the questions I raised, as the information about the electron is brought into being by the experiment that the observer chooses to make; that is, by the kind of registering equipment that he puts into place. Had I asked different questions or the same questions in a different order I would have ended up with a different word as the experimenter would have ended up with a different story for the doings of the electron. However, the power I had in bringing the particular word 'cloud' into being was partial only. A major part of the selection lay in the 'yes' and 'no' replies of the colleagues around the room. Similarly the experimenter has some substantial influence on what will happen to the electron by the choice of experiments he will do on it, 'questions he will put to nature'; but he knows there is a certain unpredictability about what any given one of his measurements will disclose, about what 'answers nature will give', about what will happen when 'God plays dice'. This comparison between the world of quantum observations and the surprise version of the game of twenty questions misses much but it makes the central point. In the game, no word is a word until that word is promoted to reality by the choice of questions asked and answers given. In the real world of quantum physics, *no elementary phenomenon is a phenomenon until it is a recorded phenomenon.*

The Copenhagen view of reality is therefore decidedly odd. It means that, *on its own* an atom or electron or whatever cannot be said to 'exist' in the full, common-sense, notion of the word.

This naturally prompts the question: 'What is an electron?' If it is not a *thing* 'out there', existing in its own right, why can we talk so confidently about electrons?

Bohr's philosophy seems to demote electrons and other quantum entities to a rather abstract status. On the other hand if we simply go ahead and apply the rules of quantum mechanics *as if* the electron were real then we still seem to get the right results; we can compute answers to all well-posed physical questions, such as how much energy does an atomic electron have, and obtain agreement with experiments.

A typical quantum calculation involving electrons consists of computing the lifetime of the excited state of an atom. If we know that the atom is excited at time t_1, then quantum mechanics enables us to compute the probability that it will no longer be excited at some later time t_2. Quantum mechanics thus provides us with an *algorithm* for relating two observations, one at t_1 and the other at t_2. The so-called 'atom' enters here as a model which enables the algorithm to predict a specific result. We never actually observe the atom directly during the decay process. All we know about it is contained in the observations of its energy at t_1 and t_2. Clearly, we do not *need* to assume anything more about the atom other than what is necessary for us to obtain satisfactory results for our predictions of actual observations. As the concept of 'atom' is only ever encountered in practice when we conduct observations on it, it could be argued that all the physicist needs to be concerned with is consistently relating the results of observations. It is unnecessary for the atom 'to exist really' as an independent thing for this consistency to be achieved. In other words, 'atom' is simply a convenient way of talking about what is nothing but a set of mathematical relations connecting different observations.

The philosophy that the reality of the world is rooted in observations is akin to what is known as logical positivism. It seems, perhaps, alien to us because, in most cases, the world still behaves *as if* it had an independent existence. It is actually only when we witness quantum phenomena that this impression looks untenable. Even then, in their practical work, many physicists continue to think of the microworld in the common-sense way.

The reason for this is that many of the purely abstract, mathematical concepts employed become so familiar that they assume a spurious air of reality in their own right. This is also true of classical physics. Take the concept of energy for example.

Energy is a purely abstract quantity, introduced into physics as a useful model with which we can short-cut complex calculations. You cannot see or touch energy, yet the word is now so much part of daily conversation that people think of energy as a tangible entity with an existence of its own. In reality, energy is merely part of a set of mathematical relationships that connect together observations of mechanical processes in a simple way. What Bohr's philosophy suggests is that words like electron, photon or atom should be regarded in the same way – as useful models that consolidate in our imagination what is actually only a set of mathematical relations connecting observations.

The paradox of measurement

Bohr's so-called Copenhagen interpretation of quantum mechanics, in spite of its strange overtones, is actually the 'official' view among professional physicists. In the practical application of quantum mechanics the physicist rarely needs to confront any epistemological problems. So long as the quantum rules are applied systematically, the theory does all that can be expected of it; that is , it correctly predicts the results of actual measurements – which is, after all, the business of physicists. Nevertheless some physicists have not been content to leave it at that, because at the heart of the Copenhagen interpretation there seems to be a devastating paradox.

Central to Bohr's view is that we can generally only speak meaningfully about the physical attributes of a quantum system after a specific measurement (or observation) has been made. Clearly this gives a crucial and special physical status to the act of measurement. As we have seen, specifying the measurement's context requires particular statements about the type and location of apparatus. Implicit in this is that we can all agree on the meaning to be attached to phrases such as 'a geiger counter placed 2 m from the source'. The trouble arises when we ask where the dividing line comes between a quantum system and a piece of macroscopic apparatus. Geiger counters are, after all, themselves made of atoms, and subject to quantum behaviour.

According to the rules of quantum mechanics, a quantum system can evolve in time in two quite distinct ways. So long as the system can be considered isolated, its temporal development is described by what mathematicians call a unitary operation. In more physical terms unitary development corresponds to something like this. Suppose the state of the system consists of several different wave patterns superimposed (see p. 7). The different component waves will continually interfere with each other and produce a complex, changing pattern, analogous to the ripples on the surface of a pond. In fact, the description of this quantum evolution is very much like that of any other wave-like system.

In contrast, suppose now that a certain sort of measurement is made. The effect is dramatic. Suddenly all but one of the contributing waves disappear, leaving only a single wave pattern corresponding to 'the answer'. Interference effects cease and the subsequent wave pattern is totally transformed. (An example of this was given on p. 21.) Such a measurement-like evolution in the wave is irreversible. We cannot undo it and restore the original complex wave pattern. Mathematically, this transition is 'non-unitary'.

How can we understand these two completely different modes of behaviour in a quantum system? Obviously, the abrupt change which occurs when a measurement takes place has something to do with the fact that the quantum system is coupled to a piece of measuring apparatus with which it interacts. It is no longer isolated. The mathematician J. von Neumann was able to prove for a model system that such a coupling will indeed have the aforementioned effects. However, we here encounter once again the fundamental paradox of measurement. The measuring apparatus is itself made of atoms and so subject to the rules of quantum behaviour. In practice we do not notice any quantum effects in macroscopic devices because such effects are so small. Nevertheless, if quantum mechanics is a consistent theory, the quantum effects must be present, however large the apparatus may be. We could then choose to regard the coupled arrangement of measured object plus measuring apparatus as a single large quantum system. But, assuming the combined system can be considered isolated from yet further systems, the same rules of quantum mechanics now apply to the larger systems, including the rule of unitary development.

Why is this a problem? Suppose that the original quantum system was in a superposition of two states. Recall, for example, the case of the polarized light at 45° to the polarizer, in which the incoming state is a superposition of two possible photon states, one parallel and the other perpendicular to the polarizer. The purpose of the measurement is to see whether the photon is passed or blocked by the polarizer. The measuring apparatus will have two macroscopic states, each correlated with the two polarization states of the photon. The trouble is that, according to the laws of quantum mechanics applied to the combined system, the apparatus now passes into a superposition of states! True, if the device is properly designed, any interference effects caused by the overlap (interference) of these two states will be miniscule. But, in principle, the effects are there, and we are forced to conclude that the apparatus is now itself in the sort of indeterminate limbo state that we have come to accept for electrons, photons, etc.

Von Neumann concluded that the measuring apparatus can only be deemed to have actually accomplished an irreversible act of measurement when it too is subjected to a measurement, and thus prompted into making up its mind' (technically called the collapse of the wave function onto a particular eigenstate). But now we fall into an infinite regress, for this second measuring device itself requires another device to 'collapse' it into a state of concrete reality, and so on. It is as though the coupling of the apparatus to a quantum system enables the ghost-like superposition of quantum states to invade the laboratory! This ability for us to put macroscopic objects into quantum superposition dramatically demonstrates the peculiarity of quantum theory.

Schrödinger's cat paradox, and worse

In 1935 Erwin Schrödinger, one of the founders of quantum mechanics, had already perceived how the philosophical problems of a quantum superposition could appear at the macroscopic level. He illustrated the issue with a touch of showmanship in a now famous thought experiment involving a cat (Fig. 9):

> A cat is penned up in a steel chamber, along with the following diabolical device (which must be secured against direct interference by the cat): in a Geiger counter there is a tiny bit of radioactive substance, so small, that perhaps in the course of one hour one of the atoms decays, but also with equal probability, perhaps none; if it happens, the counter tube discharges and through a relay releases a hammer which shatters a small flask of hydrocyanic acid. If one has left this entire system to itself for an hour, one would say that the cat lives if meanwhile no atom has decayed. The first atomic decay would have poisoned it.

In our own minds we are quite clear that the cat must be *either* dead *or* alive. On the other hand, according to the rules of quantum mechanics the total system within the box is in a *superposition* of two states, one with a live cat, the other a dead cat. But what sense can we make of a live–dead cat? Presumably

Fig. 9. The paradox of Schrödinger's cat. The poison device is a means of amplifying a quantum superposition of states to a macroscopic scale, where a paradoxical coexistence of live and dead cats seems to be implied. (From B. S. DeWitt (1970) 'Quantum mechanics and reality', Physics Today, **23**, *9.)*

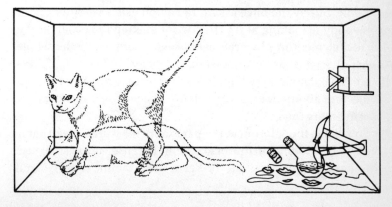

the cat itself knows whether it is alive or dead; yet, accepting the reasoning of von Neumann's regress, we are obliged to conclude that the hapless creature remains in a state of suspended animation until someone peeps into the box to check on it, at which point it is either projected into full vitality or else instantly dispatched!

The paradox becomes more acute if the cat is replaced by a person, for then the friend who has been incarcerated in the box will be aware throughout as to his health or otherwise. If the experimenter opens the box to discover that the subject is still alive, he can then ask his friend how he felt prior to this, apparently crucial, observation. Obviously the friend will reply that he remained 100% alive at all times. Yet this flies in the face of quantum mechanics, which insists that the friend is in a state of live–dead superposition before the contents of the box were inspected.

The paradox of the cat demolishes any hope we may have had that the ghostliness of the quantum is somehow confined to the shadowy microworld of the atom, and that the paradoxical nature of reality in the atomic realm is irrelevant to daily life and experience. If quantum mechanics is accepted as a correct description of all matter, this hope is clearly misplaced. Following the logic of quantum theory to its ultimate conclusion, most of the physical universe seems to dissolve away into a shadowy fantasy.

Among others, Einstein could never accept this logical extreme. Surely, he once asked, the moon exists whether or not somebody is looking at it? The idea of making the observer the pivotal element in physical reality seems contrary to the whole spirit of science as an impersonal, objective enterprise. Unless there is a concrete world 'out there' for us to experiment on and conjecture about, does not all science degenerate into a game of chasing mere images?

So what is the solution to the paradox of measurement? That is really where our contributors come in because, as we shall see, they have very different views. Let us first examine some general positions.

The pragmatic view

Most physicists do not pursue the logic of the quantum theory to the ultimate extreme. They tacitly assume that somewhere, at some level between atoms and Geiger counters, quantum physics somehow 'turns into' classical physics, in which the independent reality of tables, chairs and moons is never doubted. Bohr said that this metamorphosis required 'an irreversible act of amplification' of the quantum disturbance, leading to a macroscopically detectable result. But he left it vague as to exactly what this act entails.

Mind over matter

The key role that observations play in quantum physics inevitably leads to questions about the nature of mind and consciousness, and their relationship with matter. The fact that, once an observation has been made on a quantum system, its state (wave function) will generally change abruptly sounds akin to the idea of 'mind over matter'. It is as though the altered mental state of the experimenter when first aware of the result of the measurement somehow feeds back into the laboratory apparatus, and thence into the quantum system, to alter *its* state too. In short, the physical state acts to alter the mental state, and the mental state reacts back on the physical state.

In an earlier section it was mentioned how von Neumann envisaged an apparently unending chain of measuring devices, each 'observing' the preceding member of the series, but none ever bringing about the 'collapse' of the wave function. The chain can then only come to an end when a conscious individual is involved. Only with the entry of the result of measurement into somebody's consciousness will the entire pyramid of quantum 'limbo' states collapse into concrete reality.

Eugene Wigner is one physicist who has strongly advocated this version of events. According to Wigner, mind plays the fundamental part in bringing about the abrupt irreversible change in quantum state that characterizes a measurement. It is not enough to equip the laboratory with complicated automatic recording devices, video cameras and the like. Unless somebody actually looks to see where the pointer is on the meter (or actually watches the video record), the quantum state will remain in limbo.

In the last section we saw how Schrödinger employed a cat in his thought experiment. A cat is a macroscopic system that is sufficiently complex for two alternative states (live and dead) to be dramatically distinct. Yet is a cat complex enough to count as an observer, and irreversibly alter the quantum state (i.e. 'collapse the wave function')? And if a cat will do, how about a mouse? Or a cockroach? An amoeba? Where does consciousness first enter in the elaborate hierarchy of terrestrial life?

The foregoing considerations are closely associated with the vexed issue of the mind–body problem in philosophy. At one time many people adhered to what the philosopher Gilbert Ryle called the 'official view' of the relationship between mind and body (or brain), which can be traced back at least to Descartes. According to this view, mind (or soul) is a type of substance, a special type of ephemeral, intangible substance, different from, but coupled to, the very tangible sort of stuff of which our bodies are made. Mind, then, is a *thing* which can have states – mental states – that can be altered (by receiving sense data) as a result of its coupling to the brain. But this is not all. The link which couples brain and mind works both ways, enabling us to impress our will upon our brains, hence bodies.

Today, however, these dualistic ideas have fallen out of favour with many scientists, who prefer to regard the brain as a highly complex, but otherwise unmysterious electrochemical machine, subject to the laws of physics in the same way as any other machine. The internal states of the brain ought therefore to be entirely determined by its past states plus the effects of any incoming sense data. Similarly, the output signals from the brain, which control what we like to call 'behaviour', are equally fully determined by the internal state of the brain at the time.

The difficulty with this materialistic description of the brain is that it seems to reduce people to mere automata, allowing no room for an independent mind, or free will. If every nerve impulse is legislated by the laws of physics, how can mind intrude into its operation? But if mind does not intrude, how is it that we are apparently able to *control* our bodies according to personal volition?

With the discovery of quantum mechanics, a number of people, most notably Arthur Eddington, belived they had overcome this impasse. Because quantum systems are inherently indeterministic, the mechanistic picture of all physical systems, including the brain, is known to be false. Heisenberg's uncertainty principle usually permits a range of possible outcomes for any given physical state, and it is easy to conjecture that consciousness, or mind, could have a vote in deciding which of the available alternatives is actually realized.

Picture, then, an electron in some brain cell that is critically tuned to fire. Quantum mechanics allows the electron to roam over a range of trajectories. Perhaps it only needs the mind to load the quantum dice a little, thus prodding the electron more favourably in a certain direction, for the brain cell to fire, initiating a whole cascade of dependent electrical activity, culminating in, say, the raising of an arm.

Whatever its appeal, the idea that mind finds its expression in the world by courtesy of the quantum uncertainty principle is not really taken very seriously, not least because the electrical activity of the brain seems to be more robust than that. After all, if brain cells operate at the quantum level, the entire network is vulnerable to random maverick quantum fluctuations by any one of myriads of electrons.

The whole concept of mind being an entity capable of interaction with matter has been severely criticized as a category mistake by Ryle, who derides the 'official view' of the mind as 'the ghost in the machine'. Ryle makes the point that when we talk about the brain we employ concepts appropriate to a certain level of description. On the other hand, discussion of the mind refers to an altogether different, more abstract, level of description. It is rather like the difference between the Government and the British Constitution, the former being a concrete collection of individuals, the latter an abstract set of ideas. Ryle argues that it is as meaningless to talk about communication between the

Government and the Constitution as it is to talk about the mind communicating with the brain.

A better analogy, perhaps more suited to the modern era, can be found in the concepts of hardware and software in computing. Computer hardware plays the role of the brain, while the software is analogous to the mind. We can happily accept that the output of a computer is rigidly determined entirely by the laws of electric circuitry, plus whatever input is used. We seldom ask: 'How does the program manage to make all those little circuits fire in the right sequence?' Yet we are still happy to give an equivalent description in software language, using concepts like input, output, calculation, data, answer, etc.

The twin descriptions of hardware and software applied to the operation of computers are mutually complementary, not contradictory. The situation thus closely parallels quantum mechanics, with Bohr's principle of complementarity. Indeed, the analogy is very close indeed when we consider the question of wave–particle duality. As we have seen, a quantum wave is really a description of our *knowledge* of the system (i.e. a software concept), whereas a particle is a piece of hardware. The paradox of quantum mechanics is that somehow the hardware and software levels of description have become inextricably entangled. It seems we shall not understand the ghost in the atom until we understand the ghost in the machine.

The many-universes interpretation

So long as one is dealing with a finite system it is possible to ignore the conceptual problems associated with the quantum measurement process. One can always rely on the interaction with the wider environment to collapse the wave function. This line of reasoning fails completely, however, when we consider the subject of quantum cosmology. If we apply quantum mechanics to the whole universe, the notion of an external measuring apparatus is meaningless. Unless mind is to be involved somehow, the physicist who wishes to make sense of quantum cosmology seems to be forced to find a meaning to the act of measurement from the quantum states themselves, as it is no longer possible for an irreversible collapse of the wave function to be brought about by an external measuring device.

Interest in quantum cosmology grew in the 1960s, with the discovery of a number of theorems concerning spacetime singularities. These are rather like boundaries or edges of spacetime at which all known physics is extinguished. Singularities are formed from intense gravitational fields, and are expected to exist inside black holes. It is also believed that the universe began with a singularity. Because singularities represent the complete breakdown of physics they are considered distasteful pathologies by some physicists. It is suspected that singularities may be an artefact of our incomplete state of knowledge about gravity, which currently fails to incorporate quantum effects satisfactorily. If quantum effects could be included, it has been argued, then the singularities might go away. To abolish the big bang singularity we have to make sense of quantum cosmology.

In 1957 Hugh Everett proposed a radical alternative interpretation of quantum mechanics that removes the conceptual obstacles to quantum cosmology. Recall that the essence of the measurement problem is to understand how a quantum system which is in a superposition of, say two or more states, jumps abruptly to one particular state with a well-defined observable, as a result of a measurement (Fig. 10). A good example is the Schrödinger cat experiment discussed on p. 29. There the quantum system can evolve into two very different states: live cat and dead cat. Consequently quantum mechanical ideas fail to explain how the superposition of live–dead cat switches to either the live or dead alternative.

According to Everett the transition occurs because the universe splits into two copies, one containing a live cat and the other a dead cat. Both universes contain one copy of the experimenter too, each of whom thinks he is unique. In general, if a quantum system is in a superposition of, say, n quantum states, then, on measurement, the universe will split into n copies. In most cases, n is infinite. Hence we must accept that there are actually an infinity of 'parallel worlds' co-existing alongside the one we see at any instant. Morever, there are an infinity of individuals, more or less identical with each of us, inhabiting these worlds. It is a bizarre thought.

In the original version of the theory it was supposed that every time a measurement takes place the universe branches again, though it was always vague as to exactly what constitutes a measurement. Sometimes the words 'measurement-like interaction' were used, and it seems as though splitting can occur even from the ordinary cavorting of unobserved atoms. A proponent of the many-universes interpretation, Bryce DeWitt, expresses it as follows:

> Every quantum transition taking place in every star, in every galaxy, in every remote corner of the universe is splitting our local world on Earth into myriads of copies of itself . . . Here is schizophrenia with a vengeance.

More recently David Deutsch (see Chapter 6) has modified the theory slightly, so that the number of universes remains fixed; there is no branching. Instead, most of the universes start out

Fig. 10. The branching universe. According to Everett, when a quantum system is presented with a choice of outcomes, the universe splits so that all possible choices are realized. This implies that any given universe is continually branching into a stupendous number of near-copies.

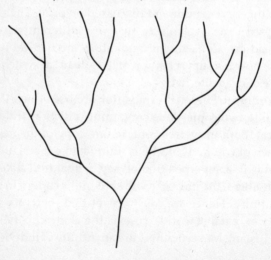

completely identical. When a measurement takes place, differentiation occurs. Thus, in the Schrödinger cat experiment, two previously identical universes differentiate so that in one the cat remains alive while in the other it dies. One advantage of this new picture is that it avoids the misleading impression that something mechanical is going on as would seem to be the case if the universe actually splits.

Two major criticisms have been levelled against the many-universes theory. The first is that it introduces a preposterous amount of 'excess metaphysical baggage' into our description of the physical world. We only ever experience one universe, so to introduce an infinity of others merely to explain a subtle technical feature (the collapse of the wave function) in our own seems to be the antithesis of Occam's Razor.

In its defence, proponents of the theory argue that the theoretical 'hardware' is less relevant in a theory than the number of fundamental assumptions you have to make in formulating the theory. The other interpretations of quantum mechanics all introduce some sort of epistemological hypothesis in order to make sense out of what is at first sight a senseless theory. The many-universes theory, however, has no need of this. The interpretation, it is claimed, emerges automatically from the formal rules of quantum mechanics, without the need for any assumptions about what the theory means. There is no need to introduce a separate postulate that, on measurement, the wave function collapses. By definition, every alternative universe contains one each of the possible collapsed wave functions.

The second objection to the theory is that it is said to be untestable. If our consciousness is confined to one universe at a time, how could we ever confirm or refute the existence of all the others? Remarkably, as we shall see, it may actually be possible to test the theory after all, if one is prepared to accept the possibility of intelligent computers.

A final argument in favour of the existence of a whole ensemble of universes is that it would provide an easy explanation for the formidable range of mysterious 'coincidences' and 'accidents of nature' found in physics, biology and cosmology. For example, it turns out that, on a large scale, the universe is ordered in a remarkable way, with matter and energy distributed in a highly improbable fashion. It is hard to explain how such a fortuitous arrangement just happened to emerge from the random chaos of the big bang. If the many-universes theory were correct, however, the seemingly contrived organization of the cosmos would be no mystery. We could safely assume that *all* possible arrangements of matter and energy are represented somewhere among the infinite ensemble of universes. Only in a minute proportion of the total would things be arranged so precisely that living organisms, hence observers, arise. Consequently, it is only that very atypical fraction that ever get observed. In short, our universe is remarkable because we have selected it by our own existence!

The statistical interpretation

In this way of looking at things, the physicist abandons all attempts to find out what actually goes on in an individual quantum measurement event and falls back instead on statements about whole collections of measurements. Quantum mechanics correctly predicts the probabilities of the various outcomes of measurement, and as long as attention is restricted to the overall statistics, there is no case to answer as regards the measurement problems.

It may be objected that the statistical (or ensemble) interpretation does not solve the problem of measurement, it simply sidesteps it. The price paid is that there is no longer any hope of discussing what actually happens when a particular measurement takes place.

The quantum potential

Another, entirely different approach has evolved out of the attempts to construct a hidden variables theory of quantum mechanics. As discussed on page 17, quantum mechanics predicts that Bell's inequality is violated. If this is correct then it is necessary to relinquish one of the two physical assumptions that went into proving it. One of these is 'reality'. As we have seen, Bohr's Copenhagen interpretation adopts this stance. The other assumption is that of 'locality': roughly speaking, there should be no propagation of physical effects faster than light.

If locality is abandoned, it is possible to re-create a description of the microworld closely similar to that of the everyday world, with objects having a concrete independent existence in well-defined states and possessing complete sets of physical attributes. No need for fuzziness now.

The trade-off is, of course, that non-local effects bring their own crop of difficulties; specifically, the ability for signals to travel backwards into the past. This would open the way to all sorts of causal paradoxes.

In spite of these difficulties, some researchers, most notably David Bohm and Basil Hiley (see Chapters 8 and 9) have pursued the idea of a non-local hidden variables theory, inventing something they call the 'quantum potential'. This is similar to the more familiar potentials associated with force fields such as gravity or electromagnetism, but differs in that the activity of the quantum potential depends on the holistic structure of the system. That is, it encodes information about the measuring apparatus, distant observers, and so on. Thus, the entire physical situation over a wide region of space (in principle the whole universe) is embodied in this potential.

In spite of all the strenuous efforts to make sense of quantum physics, there is still no unanimous agreement among physicists on the approach to adopt. Indeed, the brief survey given above by no means exhausts the full range of different interpretations that have been discussed in recent years. It is certainly remarkable that a theory which was otherwise more or less complete in its essential details half a century ago, and which has proved itself spectacularly successful in practical applications, nevertheless remains unfinished. This state of affairs is largely due to the fact that discussions of the foundations of quantum theory are largely theoretical. At best they tend to involve 'thought experiments'. The region of interest is so hard to probe that it is very rare that practical experiments can be performed to test the foundations of the theory. For this reason, Aspect's experimental test of Bell's inequality was received with enormous scientific interest.

2

Alain Aspect

Alain Aspect is an experimental physicist at the Institut d'Optique Théorique et Appliquée at Orsay in France. For several years he and his co-workers have refined the technique for carrying out a direct experimental test of Bell's inequality. The experiment reported in 1982 in *Physical Review Letters* (vol. **49**, pages 91 and 1804) was widely hailed as one of the most decisive experimental tests of the foundations of quantum mechanics ever made, and it excited great interest among theorists.

Could you briefly describe how you carried out your experiments?

It's quite difficult to describe. But we can roughly say that first we have a source which emits pairs of correlated photons, and then we have to do some kind of difficult measurements on each of these photons. Now one of the main features of our experiments was to improve the efficiency of this source. Previous attempts to study the EPR correlations led to rather uncertain results mainly because the sources employed produced only weak signals.

What source have you used?

The *ideal* source would be one atom of calcium: we would excite this atom of calcium in a particular way, and then observe the light – a pair of photons – emitted by the atom as it gives up its energy and drops back into its normal unexcited state. In fact, it's not as simple as that because we cannot trap and control a single atom of calcium so precisely. Instead, we have an atomic beam – a collection of atoms travelling in a vacuum chamber. We then excite the atoms in a very precise way by using two lasers focussed onto the atomic beam.

And this technique hadn't been used in the earlier experiments?

No, except for the experiment performed in Texas by Fry and Thompson in 1976, but there were other problems with that and their signal was not very large.

But this is only one of the new features. What other features have you introduced which improve upon earlier attempts?

From my point of view that was the main improvement, because with a stronger signal we were able to make more accurate measurements and become more confident in our results. We carried out a lot of supplementary checks to verify that everything was consistent with the predictions of quantum mechanics, and then we switched to a new kind of experiment. The second experiment we performed was closer to the original EPR thought experiment in the sense that this time we made rather more clear-cut polarization measurements. In these experiments you have to measure the polarizations of photons, the results of which can be either yes or no, either plus one or minus one. In earlier experiments as well as in our first experiment, one could only obtain directly the result plus one; the result 'minus one' of the experiment was lost. And so it was necessary to use some rather indirect reasoning for deducing what probably should have been the minus result.

So the two improvements here are a better source with better control, and the ability to be able to measure more things?

Yes.

Now, you've also been able to experiment on the pairs of photons in a way that is so rapid that there can be no communication between them at, or at less than, the speed of light. How did you do that?

Here you refer to the third experiment, and in that one we have tried to make sure that the two different parts of the system are truly independent of each other. The reason for doing this is that quantum mechanics predicts a very strong correlation between the results of the measurements on the pairs of photons even if the two sets of measuring apparatus are far from each other

(15 m in our case). One possibility for understanding this correlation in a naive picture of reality is to admit that the two
sets of measuring apparatus have some mysterious interaction
with each other. To eliminate this interpretation, some people
argue that if we rapidly change some feature, like the orientation,
of one measuring apparatus, then the other apparatus could not
respond to this change because no signal can travel faster than
the speed of light. So that's what we did.

*And this division of the two parts of the experiment into causally
disconnected regions is what is known as 'Einstein separability'.*

Yes, some people call it Einstein separability.

*And so, having performed this experiment, what conclusions do you
draw from the results?*

First I must say that this third experiment was more difficult from
a technical point of view than the previous one. So the results of
this third experiment are not so accurate. But providing they are
truly correct, we can say that the results violate Bell's inequalities,
which means that we cannot keep a simple picture of the world,
retaining Einstein' idea of separability. This is the first feature of
the results.

*So do you believe that it's possible that there is some sort of faster-
than-light signalling taking place between the separated regions?*

No, I don't think that there can be some signalling, if by signalling you mean that there is some true kind of transfer of information. What these experiments have shown is first that they
violate Bell's inequalities, and on the other hand that these
results are in very good agreement with the prediction of quantum mechanics. So we assume that quantum mechanics is still a
very good theory. Even in this kind of experiment it is not
possible to send any messages or useful information faster than
light, so I will certainly not conclude that there is faster-than-light
signalling. However, if you mean that in some picture of the
world that you want to construct, you can include some kind of
faster than light mathematical object, then perhaps, yes, it could

be a possibility. But you cannot use this mathematical construction for practical faster-than-light signalling.

What you're saying, then, is that we have to change radically the naive view of reality which most people have, to take into account this non-separability?

Yes, probably. But of course we already know that, since we knew quantum mechanics looks like a good theory, and that quantum mechanics is not compatible with the naive image of reality. But here we have shown that in this kind of very unusual situation quantum mechanics works very well, and so this must convince us that truly we must change the old picture of the world.

But some people, of course, don't like that idea. For example, the whole tradition of so-called hidden variable theories was an attempt to hold on to a naive view of reality. Do you think your experiment demolishes once and for all those hidden variable theories?

Yes, though not just these experiments (several led to a similar conclusion). However, all that they demolish is the possibility of having a hidden variable theory based on Einstein's ideas such as separability. Some hidden variable theories still remain possible: the hidden variable theories of David Bohm, for example. But note that these theories are not separable; they are not local. I mean, in these theories (such as Bohm's), there is some kind of faster than light interaction, and so we should not be surprised that these theories cannot be excluded by our experimental results.

But your experiments certainly rule out local hidden variable theories?

Yes, certainly. Provided, of course, the results remain the same in future, more sophisticated experiments.

Indeed. Are you planning any or do you know of any other group that is planning to improve upon your experiments?

No. I am not planning such an improvement because now it is

only possible to make minor technical improvements. In fact we need a new idea for making a truly important improvement, so I think that now it's enough for me.

What do you think that Einstein would have made of the result of your experiment had he been alive?

Oh, of course I cannot answer this question, but what I am sure of is that Einstein would certainly have had something very clever to say about it.

He usually did, yes!

3

John Bell

John Bell is a theoretical physicist at the Centre Européen pour la Recherche Nucléaire (CERN) near Geneva. His key theorem, proved in 1964, forms the basis for the recent experimental tests of the conceptual foundations of quantum mechanics by Aspect and others. Bell's theorem was described by Berkeley particle physicist Henry Stapp as 'the most profound discovery of science'.

Your famous result that we all know as 'Bell's inequality' can obviously only be properly discussed by using mathematics. But could you explain briefly in ordinary language what it is about?

It comes from an analysis of the consequences of the idea that there should be no action at a distance, under certain conditions that Einstein, Podolsky and Rosen focussed attention on in 1935 – conditions which lead to some very strange correlations as predicted by quantum mechanics.

By no action at a distance you mean no faster-than-light signalling?

Yes. Strictly speaking no faster-than-light signalling. In a less rigid sense no action at a distance simply means that there are no hidden connections between things.

The Nobel prize winning physicist Brian Josephson once described Bell's inequality as the most important recent advance in physics. How do you respond to that?

Well, I would say that's probably a little bit exaggerated. But if you're primarily concerned with the philosophy of physics, I can see his point.

Now, recently, it has actually been possible to put the inequality to the test rather well. One of the best experiments has been performed by Alain Aspect in Paris. What do you think of the results of this experiment? What do you think they tell us about the nature of the physical world?

Well, to begin with, one must say that the results were expected, in that they agreed with the predictions of quantum mechanics. After all, quantum mechanics is an extremely successful branch of science, and it was difficult to believe that it could be wrong. Nevertheless it was thought worth while, and I thought it worth while, to do this very particular experiment, which isolates what is one of the most peculiar features of quantum mechanics. Previously we were just relying in a way on circumstantial evidence. Quantum mechanics had never been wrong. And now we know that it will not be wrong even in these very tricky conditions.

Of course one person who was somewhat disbelieving was Einstein, and he made the famous remark that God does not play dice with the universe. Would you say that after this experiment, and after your work, you're convinced that God does indeed play dice with the universe?

No, no, by no means. But I would also like to qualify a little bit this 'God does not play dice' business. This is something which is often quoted, and which Einstein did say rather early in his career, but afterwards he was more concerned with other aspects of quantum mechanics than with the question of indeterminism. And indeed, Aspect's particular experiment tests rather those other aspects, specifically the question of no action at a distance.

You don't think it tells us anything about the determinism or indeterminism or the physical world?

To say it tells you nothing, that would be going too far. I think that it is very difficult to say that any one experiment tells you about any isolated concept. I think that it's a whole world view which is tested by an experiment, and if the experiment does not verify that world view, it is not so easy to identify just which part is suspect and has to be revised. Certainly the experiment says that Einstein's world view is not tenable.

Yes, I was going to ask whether it is still possible to maintain, in the light of experimental experience, the idea of a deterministic universe?

You know, one of the ways of understanding this business is to say that the world is super-deterministic. That not only is inanimate nature deterministic, but we, the experimenters who imagine we can choose to do one experiment rather than another, are also determined. If so, the difficulty which this experimental result creates disappears.

Free will is an illusion – that gets us out of the crisis, does it?

That's correct. In the analysis it is assumed that free will is genuine, and as a result of that one finds that the intervention of the experimenter at one point has to have consequences at a remote point, in a way that influences restricted by the finite velocity of light would not permit. If the experimenter is not free to make this intervention, if that also is determined in advance, the difficulty disappears.

Turning to this issue of the experimenter, inevitably it raises questions about mind, choice, free will and so on. Do you in fact believe that mind has a fundamental role to play in physics?

I neither believe, nor disbelieve that. I think that mind is a very important phenomenon in the universe, certainly for us. Whether it is absolutely essential to introduce it into physics at this stage, I am not sure. I think the experimental facts which are usually offered to show that we must bring the observer into quantum theory do not compel us to adopt that conclusion. The Aspect experiment is a little more tricky than the others, and I can see the logic of people who say that it goes in the direction of showing that mind is essential. It's a hypothesis that we can certainly explore, but I don't know that it's the only one.

Do you believe there are still paradoxes in the question of measurement and the role of the observer?

Yes I believe that there certainly are paradoxes. The problem of measurement and the observer is the problem of where the measurement begins and ends, and where the observer begins and ends. Consider my spectacles, for example: if I take them off now, how far away must I put them before they are part of the object rather than part of the observer? There are problems like this all the way from the retina through the optic nerve to the brain and so on. I think, that – when you analyse this language that the physicists have fallen into, that physics is about the results of observations – you find that on analysis it evaporates, and nothing very clear is being said.

So that these issues haven't been fully resolved, at least to your satisfaction?

Absolutely not. And the experiment of Aspect and the Einstein–Podolsky–Rosen correlations do not help to resolve this problem, but make it harder, because Einstein's view that behind the quantum world lies a familiar classical world was a possible (and now discredited) way of solving this measurement problem – a way of reducing the observer to an incidental role in the physical world.

Bell's inequality is, as I understand it, rooted in two assumptions: the first is what we might call objective reality – the reality of the external world, independent of our observations; the second is locality, or non-separability, or no faster-than-light signalling. Now, Aspect's experiment appears to indicate that one of these two has to go. Which of the two would you like to hang on to?

Well, you see, I don't really know. For me it's not something where I have a solution to sell! For me it's a dilemma. I think it's a deep dilemma, and the resolution of it will not be trivial; it will require a substantial change in the way we look at things. But I would say that the cheapest resolution is something like going back to relativity as it was before Einstein, when people like Lorentz and Poincaré thought that there was an aether – a preferred frame of reference – but that our measuring instruments were distorted by motion in such a way that we could not detect motion through the aether. Now, in that way you can imagine that there is a preferred frame of reference, and in this preferred frame of reference things do go faster than light. But then in other frames of reference when they seem to go not only faster than light but backwards in time, that is an optical illusion.

Well, that seems a very revolutionary approach!

Revolutionary or reactionary, make your choice. But that is certainly the cheapest solution. Behind the apparent Lorentz invariance of the phenomena, there is a deeper level which is not Lorentz invariant.

Of course the theory of relativity has a tremendous amount of experimental support, and it's hard to imagine that we can actually go back to a pre-Einstein position without contradicting some of this experimental support. Do you think it's actually possible?

Well, what is not sufficiently emphasized in textbooks, in my opinion, is that the pre-Einstein position of Lorentz and Poincaré, Larmor and Fitzgerald was perfectly coherent, and is not inconsistent with relativity theory. The idea that there is an aether, and these Fitzgerald contractions and Larmor dilations occur, and that as a result the instruments do not detect motion through the aether – that is a perfectly coherent point of view.

And it was abandoned on grounds of elegance?

Well, on the grounds of philosophy; that what is unobservable does not exist. And also on grounds of simplicity, because Einstein found that the theory was both more elegant and simpler when we left out the idea of the aether. I think that the idea of the aether should be taught to students as a pedagogical device, because I find that there are lots of problems which are solved more easily by imagining the existence of an aether. But that's another story. The reason I want to go back to the idea of an aether here is because in these EPR experiments there is the suggestion that behind the scenes something is going faster than light. Now, if all Lorentz frames are equivalent, that also means that things can go backward in time.

Yes, and that is the big problem.

It introduces great problems, paradoxes of causality and so on. And so it's precisely to avoid these that I want to say there is a real causal sequence which is defined in the aether. Now the mystery is, as with Lorentz and Poincaré, that this aether does not show up at the observational level. It is as if there is some kind of conspiracy, that something is going on behind the scenes which is not allowed to appear on the scenes. And I agree that that's extremely uncomfortable.

I'm sure Einstein would turn in his grave!

Absolutely. And that's very ironic, because it is precisely his own theory of relativity which creates difficulties for this interpretation of the quantum theory (which is in the spirit of Einstein's unconventional view of quantum mechanics).

To sum up then, you would prefer to retain the notion of objective reality and throw away one of the tenets of relativity: that signals cannot travel faster than the speed of light?

Yes. One wants to be able to take a realistic view of the world, to talk about the world as if it is really there, even when it is not being observed. I certainly believe in a world that was here before me, and will be here after me, and I believe that you are part of it! And I believe that most physicists take this point of view when they are being pushed into a corner by philosophers.

But it's always seemed to me that the practice of physics is merely creating models which describe the observations that we can make on the world, and relate them together, and we have either good models or less good models, depending on how successful they are. The idea of the world 'really existing', and our theories somehow being 'right' or 'wrong' or being approximations to this reality, I think is not a very helpful one. How do you respond to that?

Well, I do find it helpful, the idea that there is a real world there, and that our business is to try to find out about it, and that the technique for doing that is indeed to make models and to see how far we can go with them in accounting for the real world.

Do you believe that there could be an ultimate theory which would be the 'correct theory' of the universe, and would describe everything exactly?

I don't know about that, but I do believe there will be theories that are better than the ones we have, in that they describe more of the universe and connect more of it up.

So you believe that the present formulation of quantum theory, which has been so tremendously successful over the last 50 years, is still only tentative, and will be replaced at some stage in the future by a better theory?

I'm quite convinced of that: quantum theory is only a temporary expedient.

What evidence is there that quantum theory is in any way unsuccessful in explaining everything we have to explain?

Well, it does not really explain things; in fact the founding fathers of quantum mechanics rather prided themselves on giving up the idea of explanation. They were very proud that they dealt only with phenomena: they refused to look behind the phenomena, regarding that as the price one had to pay for coming to terms with nature. And it is a fact of history that the people who took that agnostic attitude towards the real world on the microphysical level were very successful. At the time it was a good thing to do. But I don't believe it will be so indefinitely. Of course, I cannot produce theorems to that effect. If you go back to, say, David Hume, who made a careful analysis of our reasons for believing things, you find that there is no good reason for believing that the sun will come up tomorrow, or that this programme will ever be broadcast. It's a habit we have, of believing that things will continue very much as they did before. However, it is a fact that this seems to be a good habit! I cannot make that a theorem, because I think Hume's analysis is sound, but nevertheless I do believe it's a good habit, to look for explanations.

So, if we think ahead to perhaps 50 years in the future, where we may

*have a theory which replaces quantum mechanics, can you see this
coming about because of continuing anxiety over the interpretational
problems we've been talking about? Or do you think that there will be
some experiment, for example something that could be performed at
CERN, such as very high energy particle collisions – exploring the
micro-microworld – that could perhaps expose an area where quantum mechanics will fail?*

Well, now you're asking me to guess. It seems to me possible that
the continuing anxiety about what quantum mechanics means
will lead to still more and more tricky experiments which will
eventually find some soft spot, some point where quantum
mechanics is actually wrong.

*So the Aspect experiment is not the ultimate experiment that can be
done to test these ideas?*

I think not. It is a very important experiment, and perhaps it
marks the point where one should stop and think for a time, but
I certainly hope it is not the end. I think that the probing of what
quantum mechanics means must continue, and in fact it will
continue, whether we agree or not that it is worth while, because
many people are sufficiently fascinated and perturbed by this
that it will go on.

*What other sort of experiments could we envisage that would test
further?*

One can point to various defects in the existing experiments,
including that of Aspect. Strictly speaking these experiments do
not demonstrate the awkward correlations. You find that the
counters that are used are too inefficient, that the geometry is
inefficient, that the ideal set-up has not been realized, and there
is an enormous extrapolation required from the experiment
which can actually be done.

*So you can envisage refinements of the present basic set-up which
will be much more convincing?*

You can envisage them, but I don't want to say that I encourage
experimenters just to go on brutally like that, making the coun-

ters more efficient and so on, because I'm inclined to believe myself that the efficiency of the counters is not the important thing.

What do you think about attempts to use superconductivity and low temperature physics to explore some of the weird quantum effects on a macroscopic scale?

They do not seem to me to be promising. I think there is a very interesting analysis by A. Leggett, who concluded that the kind of macroscopic things you see in superconductivity are rather unrelated to the kind of macroscopic things which would be embarrassing for a realistic view of the world and so on – that really they are not relevant. One tends to say, 'Oh, superconductivity shows macroscopic quantum mechanics', but not in the sense we are concerned with in Einstein–Podolsky–Rosen correlations.

And you can't imagine a more elaborate arrangement that might expose these defects in quantum mechanics?

I cannot, but I hope that's only because of my limitations. I think it is very probable that the solution to our problems will come through the back door; some person who is not addressing himself to these difficulties with which I am concerned will probably see the light. An analogy that I like is that of the fly buzzing against a window when the door is open. It can be extremely useful to stand back from your problems and just wander about for a time, and it is quite possible that those of us who are somewhat fixated on these questions will not be those who see the way through.

This is so often the way in scientific discovery isn't it?

Absolutely, and this of course is the argument for pure research, which often tends to be rather undirected.

I hope the politicians are listening! Do you see the difficulties with quantum mechanics as purely philosophical or interpretational, or do you think that there are some real experimental problems?

I think there are *professional* problems. That is to say, I'm a professional theoretical physicist and I would like to make a clean theory. And when I look at quantum mechanics I see that it's a dirty theory. The formulations of quantum mechanics that you find in the books involve dividing the world into an observer and an observed, and you are not told where that division comes – on which side of my spectacles it comes, for example – or at which end of my optic nerve. You're not told about this division between the observer and the observed. What you learn in the course of your apprenticeship is that for practical purposes it does not much matter where you put this division; that the ambiguity is at a level of precision far beyond human capability of testing. So you have a theory which is fundamentally ambiguous, but where the ambiguity involves decimal places remote from human abilities to test.

Of course Eugene Wigner has suggested that he can insert a very definite division between the observer and the observed, because he invokes the mind as a completely separate entity which is somehow coupled to the world, and he says that it's the entry into the mind of the observer that resolves the paradoxes which we've been discussing. So he's bringing the idea of a non-material mind to play a prominent part in the physical world. Do you have any sort of sympathy for that point of view?

Well, it's an idea that's worth exploring. But in my opinion, the difficulties associated with it are underestimated, simply because nobody has developed the theory beyond the talk stage. As soon as you try to put such theories down in mathematical equations, as soon as you try to make them Lorentz invariant, you get into great difficulties. For example, the interaction between the mind and the rest of the world, how does that occur? Does that occur over a finite region of space, at an instant of time? Clearly not, because that is not a Lorentz invariant concept.

By Lorentz invariant you mean that it doesn't have a consistent description for all observers depending on how they're moving?

That's correct. And the only way to get such a consistent description, if you assume the mind has access to a single point in time, is to also assume that it has access to only a single point in space.

This is the big difficulty that there has always been with mind; that it can't be located anywhere in space, and yet one presumably wants it to be located in time.

Absolutely, and yet Wigner wants somehow to couple that up into the equations of physics. It has simply not been done. It is simply talk, for the present.

There are of course a variety of other interpretations of the quantum formalism, and there is a certain amount of controversy over them. One of these is the many-universes interpretation. Do you have any strong feelings about it, for or against?

Yes, I have strong feelings against it, but I have to qualify that by saying that in this particular Einstein–Podolsky–Rosen situation there is some merit in the many-universes interpretation, in tackling the problem of how something can apparently happen far away sooner than it could without faster-than-light signalling. If, in a sense, everything happens, all choices are realized (somewhere among all the parallel universes), and no selection is made between the possible results of the experiment until later (which is what one version of the many-universes hypothesis implies), then we get over this difficulty.

But it does seem an extremely bizarre means of getting over it.

It's extremely bizarre, and for me that would already be enough reason to dislike it. The idea that there are all those other universes which we can't see is hard to swallow. But there are also technical problems with it which people usually gloss over or don't even realize when they study it. The actual point at which a branching occurs is supposed to be the point at which a measurement is made. But the point at which the measurement is made is totally obscure. The experiments at CERN for example take months and months, and at which particular second on which particular day the measurement is made and the branching occurs is perfectly obscure. So I believe that the many-universes interpretation is a kind of heuristic, simplified theory, which people have done on the backs of envelopes but haven't really thought through. When you do try to think it through it is *not* coherent.

Well, that's a very interesting and blunt response. We've been talking here about some fairly strange areas of physics; how did you first become interested in the foundations of quantum theory and in particular how did you come across your famous inequality?

Well, as a student I was very conscious of these problems: the apparent subjectivity of quantum mechanics, and this way of talking which seems to force you to bring in the observer but actually doesn't. I was, from a very early stage, convinced that it must be possible to formulate physics in a more professional way, in which this vagueness does not intrude. I actually avoided these questions for a number of years because I saw that people smarter than I had made little progress with them, and I got on with other more practical things. But then in Geneva in 1963 when I was busy with other things I met Professor Jauch at the University. He was concentrating on these issues, and in discussion with him I became determined to do something about them. One of the things that I specifically wanted to do was to see whether there was any real objection to this idea put forward long ago by de Broglie and Bohm that you could give a completely realistic account of all quantum phenomena. De Broglie had done that in 1927, and was laughed out of court in a way that I now regard as disgraceful, because his arguments were not refuted, they were simply trampled on. Bohm resurrected that theory in 1952, and was rather ignored. I thought that the theory of Bohm and de Broglie was in all ways equivalent to quantum mechanics for experimental purposes, but nevertheless was realistic and unambiguous. But it did have the remarkable feature of action-at-a-distance. You could see in the equations of that theory that when something happened at one point there were consequences immediately over the whole of space unrestricted by the velocity of light.

Did that worry you at that early stage, because of the inevitable paradoxes that would follow as a consequence?

The de Broglie–Bohm theory was developed for non-relativistic quantum mechanics only, and this instantaneous propagation of effects made it clear that that theory would have difficulties when you tried to extend it to the relativistic context.

Did you arrive at your result quickly? It's a very powerful and all-embracing result, proven in a very elegant way. Or was it something that you made tentative steps with and saw your way to the answer, and then went back and did the nice polished version?

It's a bit like the question of how long does a measurement take! How long does it take to make a discovery? Probably I got that equation into my head and out on to paper within about one weekend. But in the previous weeks I had been thinking intensely all around these questions. And in the previous years it had been at the back of my head continually. So it's really not possible to say how long it took to produce the result.

4

John Wheeler

John Wheeler was formerly Joseph Henry Professor at Princeton University, and is now Director of the Center for Theoretical Physics at the University of Texas at Austin. His research spans the fields of nuclear physics, on which he worked with Niels Bohr, gravitation, astrophysics, cosmology and quantum physics. In recent years he has produced a number of provocative and incisive arguments for the central role of the observer in quantum mechanics.

How would you sum up Bohr's contribution to the problem of squaring quantum theory with our common-sense notions of the world?

Niels Bohr was the leader in trying to understand the implications of quantum theory. It was with his help that Heisenberg worked out the uncertainty principle, and it was he who in the fall of 1927 enunciated the idea of complementarity: the idea that the investigation of one side of an experimental situation, such as the position of an electron, automatically excludes the possibility of looking at the other side, such as the momentum or velocity of the electron.

But, although Bohr's idea of complementarity did much to clarify arguments over the conceptual foundations of quantum theory, many people have had difficulty in grasping its full significance. Indeed, in Bohr's very last taped interview a few hours before his unexpected death, he singled out certain philosophers for particular criticism. He said, '. . . they have not that instinct that is important to learn something and that we must be prepared to learn something of very great importance

. . . They did not see that it (the complementarity description of quantum theory) was an objective description and that it was the only possible objective description.' That represents the centre of his thinking on quantum theory.

But surely Einstein also wanted to have an objective view of quantum theory?

Well, I think the word objective in Bohr's sense referred to the idea of dealing with what's right in front of you: the perceptions that you experience and the measurements you make, rather than Einstein's idea of the universe existing 'out there', independently of the observer.

I see. So by 'objective' Bohr probably meant rational. The complementarity description of quantum theory was, in his view, the only rational choice?

Yes, I think so.

What was your reaction to Bohr's view?

I would say that his view is what one might call battle tested. Bohr argued and discussed with everyone who had a point of view, so that in the end I would say that nobody has had a better picture of what quantum theory is and means.

But when Everett produced his many-universes interpretation for quantum theory you changed your mind for a while. Why was that?

Yes, the idea of Everett's interpretation of quantum theory was to take the so-called wave function of probability amplitude which in the past had normally been applied to an electron, an atom or a crystal and apply it instead to the whole universe. Because such a wave function includes the observer himself, this has the interesting consequence that there is no place left for the so-called act of measurement to alter the wave function. Everett's interpretation implies, for example, that when an electron has an equal chance of moving to the left or to the right the wave function splits into two branches of universe, one of which shows the electron going to the left and the observer seeing it going to the left and the other with the electron going to the right and the observer seeing it go to the right.

What attracted you to this remarkable idea?

I supported this to begin with, because it seemed to represent the logical follow-up of the formalism of quantum theory. I have changed my view on it today because there's too much metaphysical baggage being carried along with it, in the sense that every time you see this or that happening you have to envisage other universes in which I see something else happening. This is to make science into a kind of mysticism. But I also have a deeper objection: the Everett interpretation takes quantum theory in its present form as *the* currency, in terms of which everything has to be explained or understood, leaving the act of observation as a mere secondary phenomenon. In my view we need to find a different outlook in which the primary concept is to make meaning out of observation and, from that *derive* the formalism of quantum theory.

So you think that the many-universes approach may still be useful?

Yes, I think one has to work both sides of the railroad track.

But in the meantime you're siding with Bohr.

Yes. As regards the really fundamental foundations of knowledge, I cannot believe that nature has 'built in', as if by a corps of Swiss watchmakers, any machinery, equation or mathematical formalism which rigidly relates physical events separated in time. Rather I believe that these events go together in a higgledy-piggledy fashion and that what seem to be precise equations emerge in every case in a statistical way from the physics of large numbers; quantum theory in particular seems to work like that.

But do you think that quantum theory could be just an approximate theory and that there could be a better theory?

First, let me say quantum theory in an every-day context is unshakeable, unchallengeable, undefeatable – it's battle tested. In that sense it's like the second law of thermodynamics which tells us that heat flows from hot to cold. This too is battle tested – unshakeable, unchallengeable, invincible. Yet we know that the second law of thermodynamics does not go back to any equations written down at the beginning of time, not to any 'built in' machinery – not to any corps of Swiss watchmakers – but rather to the combination of a very large number of events.

It's in this sense that I feel that quantum theory likewise will some day be shown to depend on the mathematics of very large numbers. Even Einstein, who opposed quantum theory in so many ways, expressed the point of view that quantum theory would turn out to be like thermodynamics.

Both you and Bohr have referred to quantum measurement as a transition from atomic activity to knowledge or meaning via some process of irreversible amplification. Can we ever hope to find a description of how exactly this transition occurs?

To find the correct description of the building of knowledge out of measurement is a difficult enterprise in my view but extremely important. The process, I believe, has to be separated into two steps.

The first is the elementary quantum phenomenon which Bohr stressed so strongly. I try to put his point of view in this statement: 'No elementary quantum phenomenon is a phenomenon until it's brought to a close by an irreversible act of amplification by a detection such as the click of a geiger counter or the blackening of a grain of photographic emulsion.' This, as Bohr puts it, amounts to something that one person can speak about to another in plain language. Which brings us to the second aspect of this story. That is, putting the observation of quantum phenomenon to *use*. The impact of the alpha particle on a screen of zinc sulphide will create a flash which the eye can see. However, if this flash takes place on the surface of the moon there's no one around to make use of it, so that it's not used in the construction of knowledge. This is the most mysterious part of the whole story: what happens when we put something to use?

In the end I suspect we will have to depend on the work of our friends in the world of philosophy, though maybe philosophy is too important to be left to the philosophers! The construction of meaning – what meaning is – has been a central topic for study by philosophers for the past several decades and there's no single consequence of that study which better summarises the key point in my view than the statement of Follesdal – the Norwegian philosopher (a former student of Professor Donald Davison and now at Stanford University). He said that meaning is 'the joint product of all the evidence that is available to those who communicate'.

Communication is the essential idea. If I see something, but I'm not sure whether it's a dream or reality, there's hardly a better test than to check whether somebody else is aware of it and can confirm my observations. That's essential in distinguishing between reality and dreams. But how we convert this into anything empirical is quite another question. There I think we can lean on the work and findings of the great geneticist and statistician R. A. Fisher. This was way back in 1922, five years before the uncertainty principle and the modern outlook on quantum theory and in a context entirely different from quantum theory. He was studying the genetic composition of populations – the probability of grey eyes, the probability of blue eyes, the probability of brown eyes. Fisher gave up using probabilities as a way of distinguishing one population from another, and instead adopted the *square roots* of probabilities or what we call the probability *amplitudes*. In other words, he made the discovery that probability amplitude measures distinguishability.

And likewise the physicist William Wooters recognized in quantum theory that there are probability amplitudes and that the angle between two points in so-called Hilbert space (a kind of map of probability amplitudes) distinguishes between two atomic populations. Probability amplitudes thus provide a measure of the distinguishability of atomic populations. Certainly distinguishability is a central point in the establishment of what we call knowledge or meaning.

You've said that observation is a two-stage process. Just to make this absolutely clear, can you specify what is meant by an observer? For example, does a camera count as an observer?

Here we come once again to this vital distinction between the elementary quantum phenomenon – the blackening of a grain of photographic emulsion on a film or plate for example – on the one hand and the putting of that observation, that elementary quantum phenomenon, to use in the establishment of meaning. If the camera is destroyed after the picture is taken or even if I look at the picture and then a meteorite destroys me, the picture and the camera immediately afterwards, nothing has been accomplished by that picture in the establishment of meaning, although certainly the elementary quantum phenomenon itself has indeed been brought to a close.

And what exactly do you mean by an irreversible process of amplification? Does this process entail both the stages you've mentioned – the elementary quantum phenomenon and the establishment of meaning?

I would say that the blackening of a grain of photographic emulsion in the camera is the irreversible act of amplification. After all it's one single photon that does the trick and yet the grain contains an enormous number of atoms so the amplification is by an enormous factor. It's certainly irreversible because the grain is not going to turn back from black to white!

Fine. But if we turn back to the second stage of the measurement process – the establishment of knowledge – I can't help feeling that this is rather like Wigner's interpretation of quantum theory, that the translation from a quantum phenomenon to knowledge or meaning depends on the existence of conscious observers. Is that right?

Wigner speaks of the elementary quantum phenomenon as not really having happened unless it enters the consciousness of an observer. I would rather say that the phenomenon may have just happened but may not have been put to use. And it's not enough for just one observer to put it to use – you need a community.

Nevertheless, you still regard the existence of conscious observers as crucial to that second stage.

That's right. Although this word conscious is a little tricky here because one can think of animals as having brains that are so primitive that they may not be so completely aware as you or I are and, if some flash of light – some elementary quantum phenomenon – occurs which they're able to respond to in some way, then meaning has come into being even if it has involved consciousness at a pretty low level. So I would not like to put the stress on consciousness even though that is a significant element in this story.

According to Follesdal's statement, meaning is the joint product of all the evidence that is available to those who communicate. So it's the idea of communication that's important. As animals have to communicate, the establishment of meaning doesn't require the use of English!

So is there a distinction here which hinges crucially on the difference between living creatures and inanimate objects?

Yes, this is a most difficult question of where we draw the line.

Quite. To turn to a related topic, we've heard about the apparent paradox inherent in the EPR experiment. Yet another demonstration of the paradoxical nature of quantum theory seems to be offered in the so-called delayed-choice experiment. What do you make of this experiment?

The central idea is seen more simply in the split beam experiment. Light comes from a source and hits a half-silvered mirror and half goes through and half is reflected. These two beams are brought together again and can be allowed to cross each other at right angles without interacting. Further down the line there are two counters – one registers clicks for the photons which have travelled along what I may call the high road and the other counter registers clicks for photons along the low road – so that we seem to be dividing the light into photons which have quite clearly travelled one way or the other. It's a purely random business which counter clicks at any given instant.

But we can put at the place where the two beams cross each other a second half-silvered mirror oriented to bring the two beams back into coincidence to produce an interference effect. This second half-silvered mirror can be located so that we get

100% of the incoming light reaching one counter (constructive interference) while none arrives at the other counter (destructive interference). These interference effects can only be explained by saying that the light has travelled along *both* routes. [See diagram on p. 10, and accompanying discussion.]

Einstein raised an objection against such quantum-splitting experiments. How, he said, can you talk about the photon travelling along both routes of the apparatus? In the case that the second half-silvered mirror is missing it is indeed possible to say – in bad language – that the photon travels along either the upper road or the lower road. Yet, in the experiment where the second half-silvered mirror is inserted, one may say – in equally bad language – that the quantum of light has travelled along both routes!

I don't know any better illustration of this point than the famous picture by Charles Adams of the skier who comes to a tree with his pair of skis and then one sees the skier after he's passed. One track has gone on the left-hand side of the tree and the second track has gone on the right-hand side, but you don't see any revelation of how the skier has completed this miracle!

Bohr's answer was, of course, that you're dealing with the wave nature of light when it goes by both routes. That is, when the second half-silvered mirror is in place we use the wave picture and when it's out we use the particle picture. This is an example of the principle of complementarity. There's no contradiction because nature is so built that we can study one aspect of nature, or the other aspect, without any possibility of studying both aspects simultaneously.

But the new feature about the delayed choice version of this experiment is that we can wait until the light or photon (that is going to activate one of the counters) has accomplished almost all of its travel before we actually choose between the photon going by both routes or a photon going by only one of the two routes – again in bad language. We can wait until the very last minute (in reality a small fraction of a second) to decide whether we will put in the half-silvered mirror or leave it out. Therefore, it begins to look as if we ourselves, by a last-minute decision, have an influence on what a photon will do when it has already accomplished most of its doing! This seems to be in violation of any normal principle of causality.

However, in actuality, we have no contradiction. As Bohr said, we have no right to talk about what that photon is doing during its long travel from the point of entry – from the first half-silvered mirror to the point of registration. After all, no elementary quantum phenomenon is a phenomenon until it is registered. What we envisage as so definite is in fact like a Great Smoky Dragon. The tail of the dragon is sharp and clear: that is the place where the photon enters the equipment at the first half-silvered mirror. The mouth of the Dragon is quite clear: that is where the photon reaches one counter or the other. But in between we have no right to speak about what is present.

What is the relevance of Bohr's quantum phenomenon for our understanding of existence?

If we're ever going to find an element of nature that explains space and time, we surely have to find something that is deeper than space and time – something that itself has no localization in space and time. The amazing feature of the elementary quantum phenomenon – the Great Smoky Dragon – is exactly this. It is indeed something of a pure knowledge-theoretical character, an atom of information which has no localization in between the point of entry and the point of registration. This is the significance of the delayed-choice experiment.

Can the experiment actually be performed?

I'm delighted to say that Carroll Alley and his colleagues at the University of Maryland have been conducting just such an experiment. Bohr himself referred in a single sentence to the fact that it makes no difference in a quantum experiment whether we make the decision before the photon has started its travel or while it's already on its way. However, actually seeing this for real helps to clarify what an elementary quantum phenomenon is. The preliminary results of the Maryland experiment, which have been reported to me, indicate that Bohr's expectations are indeed fulfilled.

I believe that you suggested a hypothetical cosmological version of this experiment. Can you tell us about that?

Yes. The experiment that I've just been talking about is on a laboratory scale. However, there's nothing in principle to prevent the scale being some five billion light years if we use a light source such as a quasar. Fortunately there is a quasar by chance situated in space in such a way that its light comes to us by two different routes – each on a different side of an intervening galaxy which just happens to be in the same line of sight in the sky. The two beams are bent by the gravitational field of the intervening galaxy so that they converge on the eye of an observer here on Earth. This so-called gravitational lens effect provides us in principle with a means to do the delayed-choice experiment at the cosmological level, even though technically it is beyond us.

The photons reaching us start out more than five billion years ago – that is, before there was anyone on Earth. Waiting here on Earth we can today cast a die and at the very last minute decide whether we will observe an interference photon (that is a photon which has come, as we jokingly describe, 'both ways') or change our method of registration so that we will find out which way the photon has come. And yet the photon has already accomplished most of its travel by the time we make this decision. So this is delayed choice with a vengeance!

But let me insert a caveat. The very words, we have to appreciate, are wrong. It's wrong to say that the photon goes by one route or another, or by both. Picturesque though they are, those phrases are only suggestive. The elementary quantum phenomenon – the Great Smoky Dragon – is only brought to a close by an irreversible act of amplification in the counter, at our telescope. That Great Smoky Dragon had its heart back in that distant quasar. To the extent that it forms a part of what we call reality, we have to say that we ourselves have an undeniable part in shaping what we have always called the past. The past is not really the past until it has been registered. Or put another way, the past has no meaning or existence unless it exists as a record in the present.

Does that mean that we, as conscious observers, are responsible for the concrete reality of the universe?

That's probably too loosely stated. I would rather return to the concept of meaning as a joint product of all the information that is exchanged between those who communicate. And that information comes back to a set of many elementary quantum phenomena. Naturally most of these elementary quantum phenomena are individually at too low a scale of energy to be felt. But we know that many a chance yes/no adds up to the definiteness of how much. We know that the pressure of one hand on another goes back to atomic impact from the atoms on one hand to the atoms on the other. And each atomic impact in the last analysis is close to a yes/no process. So the pathway between the quantum and knowledge of meaning is a long one. The work of Hubel and Wiesel on the visual system of the brain has been enough to show that.

If we understood the workings of human or even animal brains, would that help to solve the problem of understanding the link between quantum theory and meaning?

The nature of the brain is certainly a sophisticated and challenging subject and extremely important. But I cannot believe that the physical elements for arriving at the description of the physical world are to be found there. Meaning, yes. That may depend on our unravelling the detailed mechanisms of the brain.

But it's interesting to recognize that the human brain may not be so impressive as we're accustomed to believe. The studies of the evolution of the eye have shown that the eye has independently evolved in other species over 40 times. And insofar as the eye is the window of the mind, we can believe that the evolution of the mind too is not so special as we might think.

Can you foresee any further experiments to test the conceptual foundations of quantum theory?

I am too stupid to see any immediate experiment or test. I would rather hope that we shall find a deeper conceptual foundation from which we can *derive* quantum theory; a foundation based upon distinguishability on the one hand and complementarity on the other. Complementarity restricts our freedom in the asking of questions, whilst distinguishability clarifies the result of answering these questions. But spelling out the details of any such derivation is still beyond us. So it's on this conceptual side – the derivation side – rather than the experimental side that I think the greatest hope of progress lies.

5

Rudolf Peierls

Sir Rudolf Peierls retired from the Wykeham Chair of Physics at the University of Oxford in 1974. His interest in quantum mechanics dates from his early studies under Sommerfeld, Heisenberg and Pauli, and his frequent visits to Bohr's institute in Copenhagen. He has therefore been acquainted with the 'Copenhagen interpretation' for half a century, and he still finds it satisfactory.

How did you first become interested in the conceptual foundations of quantum mechanics?

I was a graduate student when quantum mechanics started and of course it was an exciting time: together with trying to understand how to use quantum mechanics, which obviously we all wanted to do, we also had to understand what it meant.

And were you influenced by the Einstein–Bohr debate, which we read about now as a great historical episode? Is this something which had an impact on you?

No, because I learnt about it only later, not while it was taking place, but we, of course, were sure that on that particular debate Bohr was right and Einstein was wrong.

Well, I'd like to ask you about that, because as far as the textbooks would have us believe it's certainly considered these days that Bohr's Copenhagen interpretation is the official view. Yet, strangely, it seems rather difficult to find people these days who are prepared to pin their colours to the Bohr mast. Do you think that the Copenhagen interpretation is still the official view?

Well, first of all I object to the term Copenhagen interpretation.

Why is that?

Because this sounds as if there were several interpretations of quantum mechanics. There is only one. There is only one way in which you can understand quantum mechanics. There are a number of people who are unhappy about this, and are trying to find something else. But nobody has found anything else which is consistent yet, so when you refer to the Copenhagen interpretation of the mechanics what you really mean is quantum mechanics. And therefore the majority of physicists don't use the term; it's mostly used by philosophers.

So Bohr's interpretation, in your opinion, is the only one that we can really take seriously at present, although there are attempts to find alternatives. Perhaps you could tell us what you understand by the Copenhagen interpretation, if I dare call it that?

I have to say, to start with, that it is a little hard to get used to because it seems to contradict our intuition in many respects. Our intuition tells us, for example, that if you have an electron or some other particle, it is always to be found at some place with some particular velocity. Quantum mechanics, however, tells you that you have to use these concepts – position and velocity – with caution because they may not necessarily have a meaning in the context of an experimental situation. Our intuition, of course, has developed from the experience of everyday life, which is on such a different scale from the atom, that these quantum effects – these complications – are unimportant. Sometimes we meet everyday situations where concepts that seem intuitively obvious suddenly lose their meaning. The simplest is the concept of up and down, which at first has an obvious intuitive meaning, until you ask the question of whether Australia is below us or above us. You then realize that there is no answer to the question.

Now, quantum mechanics meets situations like that all the time, and therefore we have to realize that we should invoke only concepts which have a meaning inasmuch as they can be connected to some real or at least possible experiments. That gives

gives many people the trouble that they want to see *reality*; for example, they want to say 'Well, maybe I can't observe where the electron is, but *really* it ought to be somewhere'. The word 'real' here is a concept which is not clearly defined, and in my opinion all the worries about quantum theory are due to the fact that one is using terms which are not defined. Reality, of course, in everyday life is quite clear; this table we are sitting at is real because I can see it, I can touch it, and I feel a pain if I knock it; but obviously when you talk about the reality of an electron this is not what you mean.

Could I interrupt you, and ask you whether you believe that if we don't look at the table because we're in another room perhaps, whether it is still really *there?*

Oh yes. Because there are many ways in which the table's existence makes itself felt. On the everyday scale of classical physics, where an observation does not appreciably interfere with the object that's being observed, you can talk lightly about all these concepts without getting into trouble. But in quantum mechanics, it's different because any process of observation must involve an interference with the object you are observing. And therefore, in talking about what the object is doing, we must be very specific about what we observe, or what we're free to observe.

Bohr, as I understand it, expressed it this way: if we're to talk about reality, it's always within the context of a specified experimental arrangement; you've got to say precisely what you're going to measure, and how you're going to do it, before you can say what is actually going on.

That is right.

Is it correct then that we can't think of an electron as being just like a scaled down version of say a billiard ball, in the sense that we can't say that it has a position or has a motion, until we've actually measured either its position or its motion? In the absence of a measurement we can't say that it has either of these qualities?

Yes, I entirely agree with this way of putting it.

Of course this makes the external world appear rather ghostly, because it seems to require the presence of an observer before it exhibits definite attributes. Many people have drawn the conclusion that therefore mind *must play some fundamental role in physics because it's only when we're talking about observations that we can really talk about reality. Do you think mind has such a role to play in physics, or could we replace the observer with some inanimate device?*

No we couldn't. That brings in a very interesting question. In quantum mechanics we always talk in terms of what is called the wave function, or a state function of a system, and this is a mathematical object which represents our knowledge of the system; knowledge of an electron for example. Now, when we make an observation, we have to replace that wave function description by the new one which takes into account our new knowledge of the system. And there has been a lot of speculation of what's involved in this 'collapse of the wave packet' as it is sometimes called.

Which is the sudden change in the wave function that occurs when an observation is made?

That's right. Now, let's think about the way an experiment – an observation – is carried out. Suppose you have an apparatus which tells you whether or not a radioactive atom has decayed by the position of a pointer on a dial. You could describe this apparatus in terms of conventional physics. But before you look at the apparatus there are two possibilities for what the result might be, and quantum mechanics will give you the probability of the pointer being in one position or the other. And then you say, alright, you have to look at it, so you shine light on the pointer, but again you only know the probability of the light being reflected in one direction or the other. So it goes on until the moment at which you can throw away one possibility and keep only the other is when you finally become *conscious* of the fact that the experiment has given one result.

So you think consciousness plays a crucial role in the nature of reality?

I don't know what reality is.

Well, let me put it this way. Supposing that rather than getting a human being to conduct this experiment, we replace him by a very advanced computing system, or maybe even something more modest like a camera. Can the operation of a camera recording the position of the pointer on film be considered to collapse the wave function, putting that radioactive atom into a concrete condition?

No, it will not, because you can of course describe the operation of your camera or of your computer by the laws of physics, and what you will find is that when the camera has been exposed or when the computer has taken in the information, both options are still open, and so there's no collapse of the wave packet.

So the ghostliness of the micro-world gets amplified up and becomes a ghostliness of the camera or the computer?

Well, I wouldn't call it ghostliness.

Indecision?

Yes, that relates to knowledge. You see, the quantum mechanical description is in terms of knowledge. And knowledge requires *somebody* who knows.

But can a computer know?

I would say not.

So this does seem to suggest that there is a quality of human beings, call it the mind, that distinguishes us from the other objects in our environment and which is absolutely crucial for making sense of fundamental physics?

I think that's right. And, in fact, that has an interesting consequence because some people say, 'Well, supposing you include the observer in your quantum mechanical or wave function description, you could write down equations describing the motion of every electron in every cell of the brain of the observer.'

You couldn't actually do it, but in principle such equations should exist. Having established this wave function the question would be: 'whose knowledge does this represent?' There's no easy answer to that.

No, I'm sure there isn't!

I think the way out of this is simply that the premise that you can describe in terms of physics the whole function of a human being (or any other living being), including its knowledge, and its consciousness, is untenable. There is still something missing.

Well, surely one difficulty with this point of view is that there was obviously a time before there were any human beings and presumably before there were any observers of any sort. Can we think that in some sense the universe was unreal or undecided before there were any human beings around to exorcise the ghost worlds of quantum theory?

No. Because we now have some information about the origins of the world. We can see around us in the universe many traces of what happened there before. We haven't understood all of it clearly, but the information is there. We can therefore set up a description of the universe in terms of the information which is available to us.

This is a very interesting idea. You're saying that in a sense our existence as observers here and now, 15 billion years after the big bang, is in some sense responsible for the reality of that big bang because we're looking back and seeing the traces of it.

Again I object to your saying reality. I don't know what that is. The point is I'm not saying that our thinking about the universe creates it as such; only that it creates a description. If physics consists of a description of what we see or what we might see and what we will see, and if there is nobody available to observe this system, then there can be no description.

That seems reasonable. But, of course, Einstein would have disagreed, I'm sure, very strongly with what you're saying because he believed that reality was something that we just uncover by our observations. Do you think Einstein was completely wrong on this one?

I think so. Despite the enormous respect we all have for Einstein for the many things he discovered in physics, we have to accept that he was not willing to adjust himself to the implications of quantum mechanics. You see, there is no clear way in which we can define this concept of reality. Now, it is true there are many valid concepts which we cannot clearly define, and therefore it doesn't necessarily follow that reality can't have a meaning. But if we try to maintain this Einstein ideal that there must be such a thing as reality, then we get into lots of logical troubles with quantum mechanics. People have spent 60 years or so, trying very hard to find a way out of those troubles but they haven't found it. It seems to me very unlikely that such a way exists.

There does seem to be a very powerful, even emotional, appeal for believing in concrete reality, or objective reality as Einstein would have it; that is, somehow to write ourselves out of the picture. Personally, I've always found it curious that scientists should want to displace the mind or the observer from the centre of things because it seems to me appealing to have us there. Why do you think there is this restless search by many physicists to find some sort of vestige of Einstein's vision of objective reality which doesn't depend on the mind?

Well, for the reasons you've just given, except that I do not think there are so many physicists who are worried about this. I think it is only a very small minority.

Perhaps they are just vocal.

They are vocal. In fact, I was asked the other day why it is that so few people are willing to stand up and defend Bohr's view, and I didn't have an answer on the spot. But the answer is, of course, that if somebody published a paper arguing that two and two makes five, there wouldn't be many mathematicians writing papers to defend the conventional view!

Of course, one of the things that Einstein did was to invent a thought experiment which promised to throw new light on this problem. Now, recently, with the results of Aspects's experiments and other similar experiments, we've seen that these ideas can actually be tested. Do you think that the results fit into an expected pattern or do you think Aspect's experiments tell us anything new about quantum mechanics?

No, they don't. Of course it is always nice to have a theoretical prediction verified by experiment, because in the past we've had surprises. But physics is not changed by the fact that these experiments have given results which agree with the predictions of quantum mechanics. If they had come out otherwise, then we would have been in real trouble, because then we would have had to abandon at least some part of our existing scheme, and in fact it's very difficult to imagine a theory which can reproduce all those many results of quantum mechanics which have already been very accurately verified, and introduce something new with these few erudite experiments. But, fortunately, that's not what happened, for the experiments have agreed with quantum mechanics.

Now there was a time when people suggested the possibility of hidden variable theories – theories in which one could think of the indecision or indeterminacy of a quantum particle as being due to the particle being jiggled around by a set of complicated random forces that we couldn't perceive, much as in thermodynamics one has particles jiggled around by a collection of complicated forces due to molecular bombardment. Do you think that Aspect's experiments have put an end to such theories, or is there still a way in which they can survive?

Well, if people are obstinate in opposing the accepted view they can think of many new possibilities, but there is no sensible view of hidden variables which doesn't conflict with these experimental results. That was proved by John Bell, who has great merit in having established this. Prior to that there was a proof due to the mathematician von Neumann, but he made an assumption which is not really necessary. So, I think the answer is that these experiments, at least, dispose of all existing hidden variables theories, but perhaps somebody can still come up with one which is compatible with these experiments.

One possibility is to abandon the idea of locality, that is, to entertain the possibility of some sort of faster-than-light signalling, so that the events taking place at separated places can somehow conspire together simultaneously; I think Einstein referred to this as 'spooky action at a distance'. If one is prepared to entertain the possibility of such instantaneous communication, then I suppose it would be possible to retain an objective view of reality and yet still be in keeping with the results of Aspect's experiment.

It becomes a very funny view of reality if you do that. First of all, if there were any real possibility of transmitting signals instantaneously or faster than light, then of course we'd again be in very great trouble with the theory of relativity.

And it might then be possible to transmit signals backwards in time, and perhaps influence our own past with all the attendant paradoxes that that would produce.

Indeed. But of course if you think about the consequences of these new types of experiments they open up no way of transmitting signals faster than light.

But if we have correlation between separated events, it seems as though they must be conspiring in some way, and letting each other know what's going on. Can you think of an easy way of explaining why that isn't so?

The original thought experiment we are talking about, involves two particles with spin. When you measure the spin in any particular direction you get a definite result – either plus or minus. The amusing point is that if you measure the spin of one particle in the vertical direction, say, then you can predict the spin of the other side equally in the vertical direction, and if you were to measure its horizontal component then you can also predict the horizontal component of the other side. And this makes people think that by choosing whether to measure either the vertical or the horizontal component this is changing the situation on the other particle in some way. But, in fact, it doesn't. Of course if you know the answer for one particle, then you also know what's true on the other. But if you make a measurement of either vertical or horizontal spin, and do not disclose the answer, then nothing has changed for the other particle, and therefore this cannot be used to transmit a signal.

You have no control over the outcome of your particular measurement, and so you can't control the outcome of the other one; all you know is that having performed your measurement then the result of the other corresponding measurement is fixed.

Yes, but whether you perform the measurement in one direction or the other doesn't alter the state of affairs on the other side. So there is no way in which this allows rapid transmission of signals. If you think of your hidden variables then you would have to invent some variables which you can never measure, which you can in principle never know the answer to, and which somehow or other communicate over long distances but without at the same time affecting the physical situation. To me that seems to be such an unattractive view that, even if you could make it consistent with Aspect's experiments, I would prefer the present interpretation.

You make such a position seem very contrived. Could I turn to the subject of cosmology, because these days there's a great deal of interest in applying quantum theory to the entire universe. And there we run into a severe interpretational problem because if the entire universe consists of everything including observers, we have a difficulty with how a measurement of the quantum state of the whole universe can ever be made. What's your reaction to that?

Well, I think it's quite clear that this would never be possible. One difficulty is that we do most of our illustrations, most of our exercises, in quantum mechanics by thinking of a system and saying 'Right, now we have measured the state of the system completely and that's where we're going to start'. Technically, that's what we call a pure state. Now, we never meet that in any practical situation. There are always so many variables left over which could, in principle, be measured but which we don't have the time or the energy to follow up. This is similar to the situation we have in classical physics where one seldom claims to have measured everything possible. Or to statistical mechanics, where you leave the behaviour of many of the individual molecules undetermined, and, instead, consider only their average behaviour. In the universe we have this problem, only more so.

But it is possible within the context of classical physics to make sense of what the entire universe is doing. We could envisage, in principle (as indeed Laplace did), having information about all of the particles and their trajectories and in some sense predicting all future behaviour. If you try to do that sort of thing with quantum mechanics you come up against the stumbling block of including the observer.

Well, you could never predict all the future behaviour, that's in the nature of quantum mechanics, which is not deterministic. But you could in principle write down a quantum mechanical equation, a wave equation, for the whole universe.

People do, yes. But the thing is, does it mean anything?

Well, it would mean something only if you could ascertain the initial conditions, and find out what the state was at some particular time in microscopic detail.

But although one might not be able actually to gain that information is it meaningful to even think in terms of the wave function of the universe?

You can think about it.

But does it make any sense?

I think so, because there would still be some consequences you could derive from its behaviour which would be independent of the nature of the original observation that somebody might have made. Those consequences are going to be useful, so I think it is legitimate to speculate about such a wave function. But we can't actually carry this out in practice because we'll never be able to do the whole observation. And you said, when you specified that wave function of the universe, it would include all observers, and there again we come onto this question of whether biology is a part of physics in the same sense in which we now know that chemistry is ultimately a part of physics. And that is not proven. Many people tend to assume it, but it may not be true.

You mean that there may be qualitatively new features which emerge when structures become sufficiently complex?

When they become alive.

And life itself is not something that can be reduced to the properties of atoms?

I don't think anything terribly mysterious is to be expected here; it's rather like what happened to physics in the nineteenth century when, at first, scientists believed that any explanation had to involve a mechanism, and that mechanics was the whole of physics. When they met electric and magnetic phenomena, physicists tried to explain these in terms of some kind of mechanism. Maxwell even tried to do that, but then he realized, and other people realized, that this didn't make sense because electricity and magnetism were physical concepts in their own right, not contradicting but adding to and enriching mechanics. And in that sense I think we won't have finished with the fundamentals of biology until we have enriched our knowledge of physics with some new concepts. What these will be I wouldn't like to say.

Some of the people who have thought about the wave function for the whole universe have felt obliged to adopt the so-called many-universes interpretation of quantum mechanics, where one envisages that all possible quantum alternatives in some sense co-exist. What do you think about this interpretation?

That's making things unnecessarily complicated. Since we have no means of seeing or ever communicating with the other universes, why invent them? There is one way of thinking along those lines which I regard as sensible but unnecessarily high-brow. Quantum mechanics can only make predictions from given initial information. When you've made some observations, you know something about your system, and then quantum mechanics can tell you what are the chances of future experiments giving another set of results. So, in a sense, you can say that quantum mechanics can be represented as a dictionary listing all possible outcomes of all possible initial conditions. Now, if instead of 'dictionary' you simply use the term 'many universes', then we are in line with Everett and the other protagonists of the many-universes idea. In other words, there are certainly many *possibilities* allowed for by quantum mechanics and we, by our observations, find out which possibility we actually see. In the more conventional language of Everett's interpretation, what you say is that you have to make an observation to see which of the *universes* you are in. But I prefer to use the word 'possibilities' or 'dictionary of possibilities', rather than 'universes'.

6

David Deutsch

David Deutsch is a Research Fellow at the Department of Astrophysics, Oxford, and at the University of Texas at Austin. He has had a long-standing interest in the conceptual foundations of physics in general and quantum mechanics in particular. He argues here for the many-universes interpretation.

Can I begin by asking you to give a brief description of the many-universes interpretation?

The idea is that there are parallel entire universes which include all the galaxies, stars and planets, all existing at the same time, and in a certain sense in the same space. And normally not communicating with each other. But if there were no communication at all then there wouldn't be any point to our postulating the other universes. The reason why we have to postulate them is that, in experiments on a microscopic level in quantum theory, they do in fact have some influence on each other.

Before we get into that, can I just clarify that this is correct: in some sense, 'out there' there are other universes much the same as this one existing alongside ours but unconnected to it through our own time and space?

That's right.

So where are these other universes?

As I said, in a sense they are here sharing the same space and time with us. But in another sense they are 'elsewhere' because the same theory which predicts their existence also predicts that we can only detect them indirectly. We can never go there or communicate with them in any large-scale way.

Now why should we believe in such a monstrous suggestion?

I suppose the first reason is that the theory which predicts them is the simplest interpretation of quantum theory, and we believe quantum theory because of its enormous experimental success: it really has been the most successful physical theory in history.

You say it's the simplest interpretation of quantum theory, but it seems to me like a very complicated interpretation, or at least an interpretation that involves some pretty bizarre ideas. In what sense is it the simplest?

It is by far the simplest in that it involves the fewest additional assumptions beyond those which correctly predict the results of experiments. All theories in physics predict some things which are directly amenable to experiment and some which aren't. For example, our theories of the stars predict things one could measure, like how brightly they will shine, and when they're going to go supernova. But they also predict things like the temperature at the centre of the star, which we cannot measure directly. We believe these ideas, including their unobservable predictions, because they are the simplest way of explaining the things we can observe within a consistent physical theory.

Now the other interpretations of quantum theory also involve rather counter-intuitive assumptions about reality. In some of them consciousness – human consciousness – has a direct bearing on the nature of physical reality, so that nothing exists until it is observed. This is, in my view, a far more spectacular and actually unacceptable consequence of the theory than the idea of parallel universes.

So the parallel universes are cheap on assumptions but expensive on universes.

Exactly right. In physics we always try to make things cheap on assumptions.

How many of these other universes are there?

The exact number depends on the details of physical theories which we don't know well enough yet. I think it's safe to say that there is a very large, probably infinite, number of these universes. Many of them are very different from ours, but some of them differ only in some minute detail like the position of a book on a table, and are identical in every other respect.

Can you say something about how these universes come into being. Are they there all the time, or do they increase or perhaps decrease in number?

In my favourite way of looking at this, there is an infinite number of them and this number is constant; that is, there is always the same number of universes. Before a choice or decision is made, in which more than one outcome is possible, all the universes are identical, but when the choice is made, they partition themselves into two groups, and in one group one outcome happens and in the other group another outcome happens. Normally these two groups don't affect each other thenceforward, but as I have said, occasionally they do.

It is sometimes said that the many-universes interpretation is also a branching-*universes interpretation. That is, that when the world is faced with a quantum alternative it splits into all of the different alternatives which are presented to it. Your view is slightly different?*

That's right. When Hugh Everett first proposed this interpretation in 1957, this was the sort of language that he used, he spoke about branching universes. The reason was that, if there was a collection of identical universes, he preferred to speak of it as being one universe. If they were all identical and remained identical then he thought it would be pointless to speak of them as being 'many' – they would merely be a different way of describing just one universe. So when I say the universes partition themselves into two groups, Everett said that one universe splits into two universes. My way of speaking about this is to say that there's always the same number of universes, and that they repartition themselves.

Is it true that as time goes on, these universes differentiate more and more; and that we can envisage them as somehow existing in parallel and not changing in number, but changing in content?

That's right, they change in complexity, and this increase in complexity is the reflection in quantum theory of the second law of thermodynamics, which is that entropy always increases, or that there is a 'forward arrow' of time.

I wouldn't quarrel with that, but one of the things that puzzles me is that the underlying structure of quantum mechanics is symmetrical under time reversal, and I can't see why it is that the changes which we're talking about should occur in a preferred direction in time. Might we not find an equal number of other universes where the complexity is decreasing in time?

It is perfectly true that the Everett formulation of quantum theory allows universes to fuse, to use his old language, or to become identical again in the language I prefer; and *a priori* in the theory there is no reason why they shouldn't predominantly do that, instead of predominantly differentiating towards the future. Or indeed why they shouldn't do both in a haphazard way. Why there should be preferred forward time direction in the differentiation of universes is the same problem as occurs in all branches of physics in explaining why there is an arrow of time.

This problem is not solved in your theory?

No. I believe that there are promising avenues of research in quantum theory which may lead to a solution, but it's not solved yet. But remember, this is the same problem which exists in every branch of physics and it's not directly solved by the Everett interpretation, nor by anything else as yet. I should add that the coming together of universes on a small scale does indeed occur, and has in effect been observed, because every time there is an interference experiment, this provides indirect evidence of the fusion of two groups of universes into one.

That sounds like an amazing statement. Could you give us a precise example of where you consider that two universes have been observed to be fusing together.

Yes, in the classic Young's two-slit experiment in optics. What one does is to pass a very weak beam of light, one photon at a time (this can also be done with other particles nowadays) through a pair of slits in such a way that some of the properties of the photon are destroyed by its passing through either of the slits separately. That is if the photon passes through one slit then some of the information stored in it is destroyed, and also if it passes through the other slit this information is destroyed. According to quantum theory some aspect of this particle – the wave function – passes through both slits simultaneously and the information is *not* lost. Now this reminds one of the old argument: is light corpuscular in nature or is it wave-like in nature? The experiment that I'm now speaking of demonstrates the 'wave-like property of photons'. However, if one then puts a detector close to either one of the slits, one always finds that the photon is detected wholly, 100%, coming out of either one slit or the other. But the very presence of this detector prevents one from successfully operating the apparatus which would detect the wave-like nature of the motion. Now, the way that Everett interprets this is to say that the result of the observation of wave-like motion tells us that at a previous instant there were two groups of universes – and that in one group the photon passed through one slit, and in the other it passed through the other slit, but that later both these photons appeared at the same position, and from then on all the universes were the same again.

Let's just get this right. We present a particle with a choice of either going through one slit or the other, and in the Everett interpretation these represent two quite separate worlds. But if we allow the system to bring these two pathways back to overlap each other, then this is like bringing the two worlds back into fusion.

Correct. And when one observes the fused photon afterwards, it has properties which rule out the possibility of it having specifically been through either the one slit or the other.

So these worlds that we're talking about, then, although not part of our space and time, nevertheless appear to be able to communicate at the level of atoms. Is it possible to envisage ever exploring these other universes? Can we ever gain any information about them, even at the atomic level? Can we look at the properties of atoms and find out anything at all about these other universes?

To a limited extent, we do. The only experiments in which we can detect the presence of the other universes are indirect experiments – rather like detecting the temperature of the inside of the sun which is 16 million degrees – by looking at the exterior at 5000 degrees. In other words, the way we detect them is via our theory.

As for exploring these other worlds, our present theory states that this is impossible. We cannot travel to them any more than we can directly travel into the past or into the future.

But nevertheless these other universes have inhabitants that look very much like you and I?

Just as the past and future do. In fact Don Page and William Wooters recently explored this connection between the 'different universes' of the past and future and the different universes existing now side by side with us and described these on a uniform mathematical footing which shows that the past and future are just special cases of Everett's other universes.

But travel into the past would involve certain types of paradoxes – causal paradoxes – that maybe are circumvented in the case of these parallel universes. One can envisage going to one of the other universes and meeting, as it were, another copy of oneself. But it wouldn't strictly be oneself, because it would be slightly different. And you could alter the future events of such a universe without contradicting your own future in your own universe when you went back. Doesn't it in fact enable you to escape from some of these famous time travel paradoxes that science fiction writers love?

It would, if quantum theory were slightly different. The reason why quantum theory doesn't allow this in its present form is that, just as the past causes what happens in the present, in some sense, and the present causes what happens in the future, so the different parallel universes are linked by being part of a common physical object. Physical reality is the set of all the universes evolving together, like a machine in which some cogwheels are connected to other cogwheels; you cannot move one without moving the others. So the parallel universes are connected as inextricably as the universes of the past and the future.

If you went to one of these other universes and stamped on a beetle, it would have a knock-on effect in your own universe?

That's right.

So it may be more complicated than we thought in these time travel paradoxes?

Yes. One could of course speculate that with a modification to quantum theory one could travel to the past, or to other universes in the present, but since quantum theory is the only reason that we have for believing in these universes at all, it seems rather wild to change quantum theory just to make the universes behave in a slightly different and even stranger way than they do already.

You've explained part of your attraction to the many universes interpretation, but what in your opinion is wrong with the standard Copenhagen interpretation of quantum mechanics?

Well, I've mentioned that the Everett interpretation is more natural in the formal sense. But the best physical reason for adopting the Everett interpretation lies in quantum cosmology. There one tries to apply quantum theory to the universe as a whole, considering the universe as a dynamical object starting with a big bang, evolving to form galaxies and so on. Then when one tries, for example by looking in a textbook, to ask what the symbols in the quantum theory mean, how does one use the wave function of the universe and the other mathematical objects that quantum theory employs to describe reality? One reads there, 'The meaning of these mathematical objects is as follows:

first consider an observer outside the quantum system under consideration . . .' And immediately one has to stop short. Postulating an outside observer is all very well when we're talking about a laboratory: we can imagine an observer sitting outside the experimental apparatus looking at it, but when the experimental apparatus – the object being described by quantum theory – is the entire universe, it's logically inconsistent to imagine an observer sitting outside it. Therefore the standard interpretation fails. It fails completely to describe quantum cosmology. Even if we knew how to write down the theory of quantum cosmology, which is quite hard incidentally, we literally wouldn't know what the symbols meant under any interpretation other than the Everett interpretation.

Let me add that, in my experience of physicists changing their views about the interpretation of quantum theory, it is very often when they begin to consider quantum cosmology that they are finally convinced that there is no alternative to the many-universes interpretation.

If we're dealing with quantum cosmology, we're in trouble with the conventional interpretation. But with the many universes we have an interpretation which appears to face up to the problem of quantum cosmology squarely and come to grips with it. It gives us, at least in principle, a consistent way of being able to talk about the quantum behaviour of the entire universe. It therefore opens up the prospect that we could look to quantum mechanics as an explanation for the very existence of the universe. That is, to talk about the coming into being of the entire universe as some sort of quantum phenomenon. Do you think this could be correct?

Yes, although I must stress that unlike most of the other things I've been saying, this is speculative. (The other things I've been saying in my opinion are not.) I think that, just as there's a strong possibility of understanding the second law of thermodynamics by using this branching structure of the Everett interpretation, there is also a possibility of understanding something about the problem of the existence of the universe as a whole.

Now, in the many-universes interpretation, one seems to hang on to some vestige of objective reality, although it's a multiplied reality.

Yes, that's one of its main advantages.

Nevertheless, it's not necessary to introduce any subjective features, like consciousness and the mind and so on. Does this theory have anything at all to say about what an observer actually is?

No. Another advantage of the Everett interpretation is that it is not necessary to present within the theory a working model of an observer. That is, it's not necessary to state in fine detail what is the difference between an observer and any other physical system. One thing that the interpretation does shed light on, by the way, is what we mean by a measurement. There are topics in the theory of measurement which are far more easily dealt with in the Everett interpretation. But these are straightforward matters compared to the question of what is consciousness. I regard it as one of the *advantages* of the Everett interpretation that it has nothing to say about this. The interpretation will work even before we have an exact knowledge of what consciousness is all about. The other interpretations will not work properly until we understand consciousness.

But of course, for many people, one of the endearing aspects of quantum mechanics is precisely that it puts the observer back into the centre of the stage. It involves mind in a non-trivial way in the operation of the universe. And they like that, because it has a certain mystical appeal. You're banishing mind from the universe, or at least you're not making it indispensable to the operation of the universe.

Yes, this is an interesting controversy. I would actually put it entirely the other way round. I believe that it is the conventional interpretations which banish minds from the realm of physical reality.

Why do you say that?

Because in them the mind is supposed to obey physical laws different from the rest of reality, and, secondly, in all the versions of the conventional interpretation that I know of, the exact nature of this new property of the mind – this new mystical property – is not specified. It is more a hope than a real theory, that maybe one day we might find new laws to describe the mind which are just such that the conventional interpretation of quantum theory will work! In the Everett interpretation, the existing laws of physics are supposed to describe the mind properly, and until we find a contradiction, it's perfectly reasonable for us to carry on believing that. It's only in the Everett interpretation that the observer is considered to be an intrinsic part of the universe that he is measuring.

But he seems to be there just for the ride, he doesn't play a role in determining reality.

He doesn't play a *special* role in determining reality, no more than does any other physical system.

So it doesn't help us to understand what consciousness is? We can simply say that brains are more complex systems than individual atoms, and that for some unknown reason they endow the universe with consciousness.

That's right. But I don't see how it is an advantage for the interpretations competing with Everett's that they require this knowledge if they don't present it.

I think it's probably only an advantage to the mystics! So, let me put to you, then, the following: one could of course simply say that in dealing with the world, at least the world of physics, all we have are our observations; we can do experiments, we can make measurements and we try to relate them with a model; quantum mechanics provides us with an excellent model for relating the results of observations: we can just regard it as an algorithm, a method of connecting together all the things that we observe, and it works very well. So why do we need these elaborate ideas of many universes? Can't we just take quantum theory at face value?

The disadvantage of interpreting a theory purely as an instrument for predicting the results of an experiment, rather than regarding it as a true description of an objective reality, is that such a view would paralyse future progress. If I can give an analogy from an earlier era of physics: when Galileo was pressed by the inquisition to renounce his theory that the Earth moves around the sun and that this causes the apparent motion of the lights in the sky, he wasn't asked to go the whole way to saying that his theory was *false*, he was only asked to go half way. He was asked to say that although his theory was a good algorithm for predicting the positions of bright spots in the sky, he shouldn't go further and say that these spots were caused by things which actually had an objective existence as radiating material bodies in space.

Well, I wonder whether there is any real difference between these two approaches. It seems to me that in modern physics there really is no difference at all. For example, people regularly talk about virtual photons – are they really there, or aren't they really there? I don't think that question has any meaning. It seems to me all we have is a means of computing the results of different observations, and to talk about whether virtual particles are really there or not is a fruitless enterprise.

Yes. This is a slightly different sense of the term 'really there'. Whether we describe a virtual photon as a particle or a wave or as something that exists in ordinary spacetime or not is merely a difficulty in translating our physical knowledge into ordinary everyday language. But I think we do have to say that *something* is really there. If I can go back to the example of Galileo, had other physicists at that time truly been willing to accept the idea that his theory was merely an algorithm for predicting the locations of spots of light in the sky, further progress, towards Newton's theory, would have been paralysed, because although Newton's theory is one substantial intuitive step beyond Galileo's actual theory, it bears no comparison at all with the old theory of the celestial spheres. And if Newton had been content to stay with the old ontology of a celestial sphere, he would never have been able to formulate his own theory, *even as an 'instrument' or 'algorithm'*.

We have a dual reason for regarding the quantum theory as describing reality. One is because that's what we wanted the theory for in the first place, and the other is that not to do so inevitably stultifies progress.

I'm not completely convinced, because after all one could claim that the electromagnetic field is just an invention, just a word – it's not really there – and yet it hasn't impeded progress in electromagnetism.

Again, I think you're using the term 'really there' in two different senses. When we speak about electromagnetic fields, for example radio transmissions, the language that we customarily use to describe this is the language of these waves really being there; we say that they emanate from the transmitter and are received in the receiver. In fact it's quite hard to reformulate classical electromagnetic theory so that this is not so, although it is possible – one can speak entirely of the motion of the electrons in the receiver and in the transmitter without ever speaking of what transmits the influence between them. But this is a mistaken way of speaking, because had we, at Maxwell's time, forced ourselves into this way of looking at the world, subsequent developments in field theory where, for example, an energy density was ascribed to the field itself, and later the quantum theory of fields, would not have been possible.

But the field is still an abstract construct, isn't it?

It is certainly an abstract construct, but it gains its place in physics when a physical theory says that it corresponds to something real. What words we then attach to this real thing that it corresponds to, is a subsidiary matter.

But surely any model of reality in which we can have confidence, ultimately comes back to our observations? It's only at the receiving end – the observer – that one makes contact with reality, whatever sort of elaborate abstract machinery one may invent to talk about disturbances propagating and influences connecting together. I mean, ultimately, we come back to our observations – and that's all we've got isn't it? Why should we want any more?

I don't believe so. I don't believe that we would even have the observations, if they were really 'ultimately' all we had. The way we really observe things is via an intimate relationship between theory and experiment. We need both. After all, our very sense organs are the physical embodiments of certain theories; our eyes are the embodiments of certain optical theories, certain theories about colour, and three-dimensional space. One way we can tell that these are only theories is that some of them are false – some of the theories embodied in the eye are actually false theories. And when we see things, we do not rely solely on our sensory perceptions, otherwise we would never have found out that there are two types of green light, one directly green, and the other a mixture of blue and yellow.

Yes, but we only find out about them by extending the range of our capabilities through technology.

Exactly. We find out about them by extending our knowledge of the world through a combination of theory and observation: never just by observation, and never just by theory.

Well, interesting though the many-universes theory may be, is it perhaps simply a way of speaking about the world, or can it actually be tested? You've said that we can't visit these other universes, but can we devise any sort of experiment to show that they really do exist?

When Everett first put forward his interpretation, he believed that it was a pure interpretation in the technical sense of the word. In other words, that the physical predictions of quantum theory under his system were precisely identical with those under any other system. Now, I believe that this is not so, and I have recently done some work trying to elaborate the exact experimental difference between the Everett and the conventional 'interpretations'. I now have to say 'interpretations' in quotes because I believe that there are actually different formal structures for quantum theory.

So we're talking, not about two different ways of looking at the same theory, but two completely different theories?

Yes. Once I realized that at the mathematical level the two formalisms are in fact slightly different, and that there is therefore a hope in principle of constructing a crucial experimental test, my greatest difficulty in trying to think of a test was that the conventional interpretations are so woolly and imprecise that it is hard to pin down exactly what their predictions are! However, eventually I got to what I believe is the common core of all the conventional interpretations, which is this: they all say that at least by the time the result of a measurement has entered the consciousness of an observer, the wave function will have collapsed (or whatever this irreversible loss of information is called in the various versions of the conventional interpretation). Also, we know from experiment that so long as the information is still in a sub-atomic system, one which is still capable of exhibiting atomic interference, this collapse has not yet happened. So the collapse must be supposed to happen at some point in between the atomic level and the moment when an observer becomes conscious of it. Where, we don't know. The reason why we don't know is that the conventional interpretations are very vague on this point. Now, in the Everett picture this collapse of the wave function would be described as the sudden disappearance of all the universes except one.

But that of course doesn't happen?

Well, we believe it doesn't happen. But what we want is an *experiment* which will detect whether it happens or not. Here's how it works: we first find a situation in which the conventional interpretation predicts that all the other universes suddenly disappear, and where the Everett interpretation predicts that they don't disappear but they're all there in parallel. Then we find some observable consequence of their subsequently interacting with each other in an interference experiment. And we then observe one result if the Everett interpretation is true, and another result if any of the conventional interpretations is true. Simple as that.

Unfortunately, this experiment requires the observation of interference effects between two different states of an observer's memory. The reason why it has to be an observer's memory rather than just any old physical system, is not Everett's fault. It is that the conventional interpretation makes a special reference to observers as being different. The way in which these interpretations differ from the Everett interpretation is that they say that observers obey different physical laws, and Everett says they obey the same physical laws. So the place where we would expect a crucial experimental test is with quantum effects inside an observer's brain.

We're talking about quantum memory?

We're talking about quantum memory, and presumably electronic artificial intelligence.

Because our own brains really work at a classical level rather than a quantum level?

That's right. As far as we know. There are theories that they don't, but whether they do or not, it seems unlikely that we shall get control over the electronic functions of the human brain at such a fine level. Whereas, when it comes to electronic components, it's already commonplace to use some of their quantum properties; every microchip works on those principles. But even microchips are at present too crude for interference phenomena to be observed in them.

But we can envisage building some sort of artificial superbrain with a memory at the quantum level, and ask it to carry out this experiment for us, and tell us what it feels?

That's right. And it could record the results of this experiment in any way we like. It could perhaps write them down, or tell us the results; the difference – rather like in Aspect's experiments – between quantum theory and the rivals is not a matter of a small percentage, it is an all or nothing thing. In the experiment I describe, one would observe a certain atomic spin, and if it was pointing one way, Everett's interpretation would be true, and if it was pointing the other way, the conventional interpretation would be true.

Now, you've explained how one might construct this superbrain to play the role of an observer who has a quantum memory, but can you just tell us exactly what it is he's going to observe? Exactly what experiment does he perform, if we can call him he?

Yes. The experiment hinges on observing an interference phenomenon inside the mind of this artificial observer. This can either be done by someone else looking inside him, or more elegantly, by his trying to remember various things so that he can conduct an experiment on his own brain while it's working.

He can observe himself?

He can observe part of himself, yes. And what he tries to observe is an intereference phenomenon between different states of his own brain. In other words, he tries to observe the effect of different internal states of his brain in different universes interacting with each other.

How would these different internal states be set up?

They are set up in the first instance by a special sense organ which is essentially just another quantum memory unit. This sense organ is used to observe the state of an atomic system – a system with two possible states, such as an atomic spin for example. Now, quantum theory predicts that, having observed this atomic system, the observer's mind will differentiate itself into two universe branches.

So, we have an atomic system with two possible states, each of which would trigger the brain of this artificial observer to be in either one state or the other. And according to the Everett interpretation you're saying that these two brain states somehow co-exist – or at least they exist in parallel universes. But we don't let the universes get out of touch with each other. We bring them back to overlap, to interfere with each other, and this poor observer is, as it were, schizophrenic and observing both of these possibilities at once.

That's right. In effect he is feeling himself split into two copies.

And he feels himself merge again?

Yes, in effect. Of course, *we* don't have sense organs of this type, so it's hard to say what this would feel like, but when this observer exists we can ask him!

It sounds most uncomfortable!

Perhaps it will be, but then presumably he'll be a physicist so he'll enjoy doing this experiment!

How exactly will he go about it?

At the intermediate stage he will write down an affidavit to the effect that 'I am hereby observing one and only one of the two possibilities'.

And what he writes down will be different in the two different universes?

No. What he writes down will be the same in the two different universes because he won't actually say which of the two possibilities he observes. Instead, he will write: 'So that this experiment can be continued, I will not actually say which of the two I am observing but I do certify that I'm observing only one of the possibilities.' He can then continue with an interference experiment between the two parallel universes containing the different brain states and he should get a result which is compatible only with the presence of both of these brain states in his past. So, if interference occurs he can infer that these two possibilities must have existed in parallel in the past – supporting the Everett interpretation. However, if the conventional interpretation is true, then at some time during his deliberations all the universes but one will have disappeared. And although it'll still be true that he will write down 'I am observing only one', by the time he gets on to the interference phenomenon it won't work (i.e. the interference won't occur). And so he will have demonstrated that the Everett interpretation is false.

Because by his having this certain knowledge of a particular outcome he will have completely modified the wave properties of the system, and therefore altered the subsequent quantum development of the system, which could be checked by subsequent measurements?

Yes, he either will or won't, and if he has altered them in that way then the conventional interpretation is true, and if he hasn't then the Everett interpretation is true.

Which means that, in the Everett interpretation, it's possible for the observer to make up his mind, but he's got two minds.

Yes.

He's in two minds about it! When the experiment is complete, and this machine observer is asked to remember what it was he observed even though he didn't write it down at the time, what will he remember? Will he remember both?

No, he will remember neither in fact. It is a necessary consequence of the other things he does, that he must wipe out the memory of which one of those two possibilities he observed.

But he still has the memory that he only observed one of the two?

Yes, this is the key feature of my experiment: the memory that he knew one and only one of these possibilities can be retained even though he is obliged to forget which one.

You're saying he can deduce that he must have been split, because he knows that the outcome involves both possibilities co-existing?

Exactly.

If it's true that all these other universes which exist around us can couple to our universe at the atomic level, why don't we feel their presence?

In principle we could. There's no fundamental reason why we don't. It's only that our brains are sufficiently large to operate on essentially a classical level. If we had fine enough senses, then, rather like the mechanical observer in my thought experiment, we could detect or feel (whatever that would mean) the presence of the other universes.

You mean, if we could feel all the atoms creeping about in our brains, then we would indeed feel these other universes?

Yes. In fact, as I said, Everett often compared critics of his interpretation to the opponents of Galileo, who said that they did not feel the Earth move beneath them. The point being that Galileo's very theory predicted that one doesn't feel the Earth move, unless one uses sufficiently fine apparatus. And just as with a Foucault pendulum or with sufficiently delicate astronomical measurements one can detect – one can, in effect, 'feel' – the motion of the Earth, so with sufficiently fine senses we would indeed feel the presence of the other universes.

Anyway, to do the test you've just described we would need this supercomputer to tell us that the Everett interpretation is or is not correct.

Unfortunately yes. And it seems rather a long way beyond present technology to construct such a computer. Although when I say a long way I don't mean millions of years away, I mean decades away.

Well, it's certainly fascinating that there might be some prospect of actually testing these ideas within the foreseeable future. But why is it that Everett overlooked this possibility?

Now, I've never thought about that! Perhaps one of the reasons was that he had an *additional* idea connected with quantum theory, which was that its interpretation ought to follow directly from the formalism. That is, if you write down the mathematical rules of quantum theory, he thought there ought to be only one way of interpreting these. This is an extremely strong assumption to make, and it would, had it been true, have been the first physical theory in history ever to have this strong property. He hoped it was true, and therefore I think he concentrated on the similarities between the predictions of his theory and those of the rivals, thereby highlighting the fact that the rival conventional interpretation requires additional metaphysical paraphernalia whereas his doesn't. So he said, 'I take the pure formalism, and I add nothing, and I obtain my interpretation. In contrast they (the supporters of the conventional interpretation) have to add all this stuff about consciousness and so on.' Now, I think

Everett was slightly wrong. I think that even in his interpretation, one requires a little bit of extra structure in order to arrive at the interpretation. But not much – very much less than in the conventional interpretation.

Can you summarize in a few words what this little bit of extra structure is?

Yes. It is the little piece of mathematics which provides the connection between the wave function or state vector, which is the mathematical object describing the universe, and the concept of the many parallel universes. I don't think one can do without this extra structure. But I do agree with Everett so far as to say that his is the *simplest possible* addition to the purely instrumental quantum theory.

I'm not sure if I've understood this correctly. Are you saying that Everett's extra assumption is necessary to tell us something about how any individual universe in this vast stack of cosmic alternatives fits into the stack?

That's right, yes.

You have already explained the advantages that the many-universes theory has over, say, the conventional Copenhagen interpretation. What advantages do you think it has over the other rival interpretations?

There again, they are rather diverse when it comes down to details. I suppose you're referring mostly to the hidden variables interpretations?

Yes, or it's modern variant: the so-called quantum potential.

Yes. One objection is that to append to the quantum formalism an additional structure which is supposed to correspond to physical reality (this additional structure being far more complicated than the original physical theory), solely for the purpose of interpretation, is I think a very dangerous thing to do in physics. These structures are being introduced solely to solve the interpretational problems, without any physical motivation. From the point of view of a physicist, I'd say the chances of a theory which was formulated for such a reason being right are extremely remote.

But aren't you introducing many universes for precisely that reason – to solve the interpretational problem?

Well the problem of having an interpretation in the first place is itself an unavoidable problem, and I would gladly dispense with the many universes were there a simpler interpretational assumption. But the many-universes assumption is in fact so simple from the point of view of the underlying physical laws, that, as I said earlier, Everett, DeWitt and others were misled into thinking that there was no additional structure at all. It really is the most natural interpretation of the formalism yet thought of. By contrast, the hidden variables theories are very complicated. One of the reasons why they are is that from Bell's theorem and Aspect's experiments we know that the simplest forms of hidden variable theories simply cannot mock up the effect of quantum theory.

Instead, we need to have some sort of non-local hidden variable theory, which is what Bohm and Hiley are attempting to do.

A non-local hidden variable theory means, in ordinary language, a theory in which influences propagate across space and time without passing through the space in between.

Without passing through? Or should we simply say they propagate instantaneously? Is that perhaps the same thing?

Yes. To say they propagate instantaneously in the context of relativity means that they cannot be passing through the intervening space-time because if they did their description would then be inconsistent with relativity.

They don't deny this. They say of course their description is inconsistent with relativity, but when it comes to actually making measurements the results of all these measurements are consistent with relativity. It's only the mechanism itself which seems to contradict the spirit of relativity.

Yes. This is a defence only if you're willing to back completely into the corner of saying that quantum theory is merely an instrument. And if it is merely an instrument, then the hidden variable theories lose their main advantage, which is that they cling to the notion of objective reality, just as Everett does.

But look, one feature which the many-universes interpretation and these non-local, or let's say faster-than-light, interpretations have in common is that they are both attempting to retain some vestige of objective reality. In both cases, according to Bell's inequality, and the Aspect experiments, we have to make a choice. You either have faster-than-light signalling, or you throw away objective reality. Now, in my opinion, it doesn't seem terribly awful to have to throw away objective reality. Why should we insist so much that the external universe is independent of our observations? Surely it's not surprising that we ourselves play a part in reality, because we seem to be very important to ourselves? At least it doesn't surprise me, on the basis of my personal experience, that we are playing a part in reality. So why this desperate urge to cling on to some vestige of objective reality, if it means introducing complicated things like faster-than-light signalling, or other universes?

Well, I agree that Aspect's experiment forces us to change our view of reality. The reason why I would want to cling on to the philosophical notion of objective reality in itself, regardless of whether this looks unfamiliar or not is the same reason that I mentioned before for not changing to an instrumentalist view of physical theories. Firstly, because if we *can* have a theory that has objective reality in it, it is philosophically superior. Therefore we should at least try that approach before we throw away the notion of reality. And secondly, from the point of view of science, especially physics, I believe that changing to an instrumentalist interpretation of a theory makes it impossible to obtain the next theory, because the next theory will be a step forward from the ontology of our present theories. The likelihood is that it'll be even wilder, it'll tell us that the universe is *even stranger* than Everett said it was, not less strange. And if we abandon the notion of reality, we are depriving ourselves of the mechanism by which we construct conceptual models of the universe. It's only by altering our present conceptual models that we will discover the new theory.

I'm not saying we should abandon reality, but abandon a type of reality which is independent of ourselves. It just means that future models will have to incorporate the observer at a fundamental level.

Yes. I wouldn't object to that in principle. But I don't believe that quantum theory drives us to this. Perhaps I can stress again that the conventional interpretations of quantum theory which do try to give the observer a special place in forming reality haven't actually done so yet. They merely claim that they will one day.

Yes, of course. And they can't cope with quantum cosmology without having an observer outside the universe.

Yes. One day, perhaps, somebody will be able to write down exactly what physical laws the mind does obey, if it doesn't obey quantum theory. And perhaps that new physical theory (it would not be quantum theory any more, it would be a new physical theory) might be testable against quantum theory.

Well, perhaps, but nobody's written it down yet!

No, and when speaking about the supposed advantages of the conventional interpretation – that it gives the observer a fundamental view which is perhaps philosophically attractive to you (I don't know whether it is or not) – you are forgetting that in fact they don't do this yet. This is merely a claim, a promise, which over 50 years has not been fulfilled. Whereas the Everett interpretation is unproblematic. It works perfectly well without making these promises.

7

John Taylor

John G. Taylor is Professor of Mathematics at King's College, Universtity of London, and the author of a number of books, both specialist and popular. His main research interest is quantum gravity, but he is also interested in the physics of the brain. In this interview he adopts a hard-headed approach to the more outlandish ideas of quantum mechanics, and opts strongly for the statistical interpretation.

What is the ensemble (or statistical) interpretation?

Well it is a concept that lives up to its name: when we're making a measurement of any observable in a system what we're actually doing, according to the ensemble interpretation, is that we're making a measurement on an aggregate or *ensemble* of identically prepared systems. We thereby obtain a whole set of measurements, one for each of the identical versions of our particular experiment in the ensemble. Hence our results take the form of a probability distribution of particular values for that measurement.

So you just look at the statistics, and you don't care about any individual event?

That's right. And indeed it's amazing that Einstein, if I can quote him now, really did in the end settle for this ensemble interpretation. In his writings, in reply to criticisms, he said: 'One arrives at very implausible theoretical conceptions if one attempts to maintain the thesis that the statistical quantum theory is in principle capable of producing a complete description of an individual system. On the other hand these difficulties of theoretical interpretation disappear if one views the quantum mechanical description as the description of ensembles of systems.' So Einstein was, in fact, one of the forerunners of what I think is regarded by most physicists as a natural interpretation of quantum mechanical measurements. This is that we are actually doing large numbers of measurements on identical systems and we take the frequencies of particular measurement values as the probability distribution of those values.

So you make no attempt at all to describe what is going on in an individual system?

We're not allowed to. That is quite clear when we look at the various paradoxes. If we take the Einstein–Podolsky–Rosen experiment, which really is the basis of the Aspect experiment, it's clear that a paradox arises there, because we're assuming that when a measurement is made, say, of the spin of a particular particle, we can also measure the spin of a far away particle whose properties are correlated according to the usual quantum mechanical ideas. For example, we might find that the particle nearby has a spin pointing up. From that, we can conclude that the other particle far away must (if it is correlated) therefore have spin down. This would be paradoxical if you believe that you are indeed measuring individual systems because it would seem that you're actually able to influence that far away particle, and in some ways determine its spin simply by making a measurement on the nearby particle.

The ensemble interpretation says, however, that we're looking at a whole ensemble of such systems. Some 50% of them may have (when we're measuring them) nearby particles with spin up and far away ones with spin down, while the other 50% have the opposite spins. But we can't say in any particular case what that spin of the far away particle is from the measurement nearby, because we don't know about it; we only know about ensembles of such situations.

Could I press you further on this question of whether, in the ensemble interpretation, one is still supposed to think of an individual system as actually possessing well-defined properties. For example, does an

*electron at any given moment actually possess a well-defined position
and a well-defined momentum, even though of course we can't
measure what they are?*

The answer is no, the electron cannot have both of those attributes. All we can ever measure, through the uncertainty principle, are the lower bounds on the dispersion of measurements of
that position, and of that velocity or momentum over an ensemble. We can never measure these quantities for a particular
electron. It's ruled out of court, and I think we've got to accept
that from Aspect's experiment.

*But if electrons or atoms don't actually possess these properties before
we make a measurement, it seems to suggest that the observer must
therefore be involved in a fundamental way, because these particles
certainly do have well-defined properties after we make the appropriate measurements, and we of course can choose which measurements
to make, either position or momentum.*

Yes, but we're making that by setting up an ensemble: a set of
identical copies of the particular situation we're measuring.

*But we might not. We might choose to look at an electron and, say,
measure its position, and find a position, and then that's quite
satisfactory. But if we argue that it didn't have a well-defined position
before we made that measurement, then the measurement itself is
playing a crucial role.*

One has to be very careful in distinguishing between measuring
and preparing. That's something which some physicists have
considered quite carefully.

*Could you give us a brief introduction to the distinction between
them?*

Well, if you're preparing a state of an ensemble, then you know
that it will have properties identical with that preparation *in the
future.* If you make a measurement, then you will have been able
to gather what it was like just before the measurement *in the past.*
There is a rigid distinction between these two. And I think you

have to be very careful that you don't fall into the trap of always equating the measurement process with the preparation process. Once you've prepared a system, then you can begin to look at what the ensemble of states you have prepared will look like. For example, you might choose to measure the positions of a set of electrons. On the other hand, you may wish to measure their momenta. But always the dispersion of these measurements is related by the uncertainty principle. If you're going to prepare the electron in a given position, then you know that through its dispersion in the ensemble it won't have any definite value for its momentum. And that's the nature of the beast.

So you don't believe that if, for example, we prepare a quantum state with an electron at a particular position, it actually has a well-defined momentum, even though we can't ourselves measure it?

No. We must accept that it's given by all possible ranges of momenta. In other words the momentum cannot even be defined.

Yes, but that brings me back to the feeling that if its momentum can't be defined and yet after a measurement by a person it will have a well-defined momentum, then it seems that the act of measurement itself is absolutely crucial in promoting the system from a sort of fuzzy indeterminate state to one of concrete reality.

Ah, but you've then reprepared the system if you want to look at it in a state of given momentum.

But if you put it in a state of given position, and then decide to measure the momentum, of course you get a particular value, although the value can't be predicted.

Ah, but again you're making it sound as if you're looking at an individual electron.

But in practice we can do that; we can decide to make a measurement on an individual electron.

Yes, but then you will know that if you try to measure its momentum, there will be an infinite range of possibilities. There will, of course, be a particular value for a particular case in the ensemble.

Which seems to make it look like the observer intruding.

Of course. But you know very well that the preparation you make when you put the electron in a particular place gives you an ensemble in which the momentum is completely undefined. If you wish now to look at a particular case of measuring that momentum, you will get a particular value, but that value has no meaning at all in quantum mechanics. In effect, you're preparing another ensemble (if you make a number of such measurements), and if you wish to start again and say I look at all those electrons with a given momentum, now they are in no definite place.

So in this scheme the electron's wave function doesn't collapse on to one of particular momentum when you measure its momentum.

No. You're now setting up a new ensemble. You can't take a particular electron, at a given place, and say I'm now measuring its momentum, because that doesn't mean anything. That's not allowed.

If you abandon any attempt to describe what is going on in an individual system, isn't that a bit of a cop-out?

Well, I think you have to ask whether indeed you're in more trouble if you cop-in, than if you cop-out! And as far as this paradox is concerned – the Einstein–Podolski–Rosen paradox – clearly you're in great trouble. Likewise if you take the Schrödinger cat paradox. This too depends on a thought experiment.

According to any interpretation of quantum mechanics which attempts to describe individual systems, the wave function of the system *including* the cat must show that after about one lifetime of radioactive decay, there is an equal probability of the cat being alive or dead. That means the quantum mechanical state is composed of the cat being alive for half the time, and the cat being dead for the other half. In other words the cat doesn't know whether it's dead or alive, which is absolutely absurd! Now if you take the ensemble interpretation, then in 50% of the cases the cat is alive and 50% it's dead. That's quite reasonable.

So if we take one individual case, and ask whether the cat is alive or dead, then the answer – or your answer – would be that there is no answer?

Well, the answer would be that in fact, according to quantum mechanics, there is no way of saying whether it is alive or dead in any particular case. It's a meaningless question. We can only say it has a 50–50 chance of being dead (or alive). I think we have to accept that feature, especially now if we go to the Aspect experiment. Because there we see that quantum mechanics agrees with the results of that experiment and any other interpretation is not satisfactory. Except possibly for a non-local type of explanation (such as the interpretation of Bohm and Hiley). But then you have to be very careful because there are new features that enter.

If you're looking for alternative versions of quantum theory which agree with the results of the Aspect experiment, these alternatives must also stand up to the level of success we have obtained when we have gone beyond quantum mechanics to what is called quantum field theory. Quantum field theory is a whole new bag of tricks; it involves levels of success in explaining what we see in nature to one part in at least a million if not more. There are whole regimes of successes which are almost impossible to conceive of as being explained any other way.

If you think, for example, of the problems of quantum electrodynamics: one of the great successes in the late 1940s and the early 1950s was of understanding why it is that there are very delicate shifts of energy levels in the hydrogen atom which are not explained in conventional quantum mechanical terms. These energy level shifts could only be explained in terms of 'virtual' processes involving virtual photons, virtual electrons and positrons – virtual meaning not actually existing in our real world because we can't observe these virtual particles directly. Nevertheless, quantum field theory predicts very precisely the effect of these virtual processes and the results agree with the observed energy shifts to at least one part in a million. How you're going to replicate that with alternatives to quantum mechanics, my mind boggles.

To continue this train of thought, let's turn to the recent discovery of the W and Z particles – the intermediate vector bosons. These particles were predicted by a theory which unifies electromagnetism and radioactivity – a theory that was a direct product of our quantum field theories. Only by considering the implications of these quantum field theories would we have been led to the existence of these particles and to a prediction of their masses, all of which has been confirmed by high energy particle experiments at CERN. To say that any alternative to quantum mechanics could have done this is, I think, pie in the sky.

Then there are problems which I think are even more fundamental; not questions of one part in a million but questions of principle. For example, classical mechanics cannot describe the annihilation or the creation of particles. And yet we observe this all the time in particle accelerators. How on Earth is anybody going to describe this in classical terms? No amount of non-local quantum potentials or what have you will explain how matter can be created or destroyed.

So you're saying this impressive refinement of quantum mechanics called quantum field theory, which gives a very satisfactory description of large areas of modern particle physics, would simply collapse if we didn't retain the traditional notion of quantum mechanics?

Yes, I would say that any attempt to replace the uncertain quantum mechanical observables by certain but uncontrollable or hidden ones is doomed to failure. I've known physicists who, during their careers, have attempted to replace these amazing successes of quantum field theory by the classical approach. Several people spring to mind. They have all failed. And their failure has become worse and worse as the successes of quantum field theory have grown. And yet at the same time we've seen Nobel prizes awarded to some of our colleagues for their successes with quantum field theory and in particular for unifying the forces of nature. It's now very hard for me to see any other way. This avenue of research is almost unique.

Following on from these remarks one can conclude that Aspect's experiment need not have been performed because confirmation of quantum mechanics was virtually guaranteed by the enormous successes we've had with the theory so far. If you think of the understanding of locality we have gained in terms of the so-called dispersion relations based on quantum field theory, again the Aspect experiment needn't have been carried out. High energy scattering experiments have verified that locality is preserved down to billionths of a centimetre, right inside the photon. It's just impossible to conceive of any violation.

Aspect's results came as no surprise to you then?

In a way, no. Of course, there could have been a surprise round the corner, but I think I could go back to Einstein, who said that the Lord is subtle but not malicious.

Could I take you back to the Schrödinger cat paradox and ask you whether, in this ensemble interpretation, the cat is actually alive or dead but it's just that we can't know about it? Should one think of the cat being alive or dead in a particular case, even though we can't ever find the answer?

Well, we can always record the answer. And the cat itself knows whether it's alive or dead. I would have thought that the only way of avoiding a paradox here is to say that we're not allowed to find out, in any individual case. It comes, I think, to this question of the nature of consciousness. Is consciousness important in the measuring process in quantum mechanics? I think it's something that a number of physicists have claimed is a crucial feature.

Yes. Do you believe the observer is involved in the measurement process in a fundamental way?

No, because it would seem to me that we can just as well observe by means of machines, cameras, video tape recorders, and the recording equipment running here in this particular programme! I don't see consciousness as relevant at all.

I think this maybe brings us on to the question of how quantum mechanics can ever have been misused for explaining extrasensory perception, for explaining the phenomena associated with Uri Geller, spoon bending, telepathy, precognition, and all those paranormal events which of course have great public interest – great interest, for example, from the point of view of our own survival after death. All of this is related to the question of whether consciousness plays a role in fundamental physical phenomena. If consciousness is important, then maybe we can use our minds to control some very delicate physical processes, and hence explain how psychokinesis, spoon bending, and other peculiar phenomena might occur. If consciousness is not relevant then this possible connection seems to be cut.

Arthur Koestler argued in his book *The Roots of Coincidence**
that, because quantum mechanics seems to have these bizarre features associated with the Einstein–Podolsky–Rosen experiment and the Schrödinger cat paradox, that therefore other bizarre phenomena can also occur in the world. This is, I think, a very dangerous, specious argument.

Guilt by association?

Yes. But of course, with the remarkable successes that we've had in high energy physics that I have already described there is little evidence for any bizarre phenomena. High energy physics is a very precise, watertight area to work in. And, moreover I would say that there is no hard evidence at all for extrasensory perception.

A lot of people have been very impressed by how the spirit of modern quantum theory appears to be rather in tune with ancient oriental mysticism. So, quite apart from paranormal phenomena, do you regard ideas of mysticism as being of any value in modern physics?

No, I don't at all, in fact I'm rather horrified by these developments. It seems to me that there's a large amount of vague and woolly thinking that is contained in Eastern mysticism. No matter how modern science had developed, the mystics could have said, 'Aha, I told you so!' It's rather like doing the same

* Hutchinson, London, 1972.

exercise on the Bible, and picking out certain words and saying, 'Aha!, this contains all the works of James Joyce!' It's absolutely ludicrous. The detailed precision of modern theoretical physics surpasses anything that is continued in Eastern mysticism. On the other hand if these mystical ideas are used as an entrée into modern physics then they may have value, but only if used as stepping stones to the greater precision of the real thing.

Fine. Now you've said that you don't see consciousness as relevant to quantum theory but nevertheless there are a number of contending interpretations of quantum mechanics in which consciousness is involved in a fundamental way, Wigner's interpretation for example. There are also other types of interpretation such as the many universes interpretation. Now Aspect's experiments don't actually rule out these alternative interpretations because they're meant to be pure interpretations and hence be consistent with all of the known results of quantum theory. What's more they also attempt to account for what happens in individual cases. In other words, they seem to go beyond what the ensemble interpretation can do, providing more complete information about a system and getting to grips with those paradoxes. So what's your reaction to that?

Well, if it is right that they do satisfactorily come to grips with the paradoxes then I would be delighted, but I don't believe that. I am very doubtful about the consciousness interpretation mainly because it involves one in an infinite regress. Also I can't see why consciousness is so special because all it requires is an aggregate of nerve cells. In fact, consciousness involves such large aggregates of cells that it would be difficult to see how quantum effects, which involve uncertainties in rather small objects, could be significant.

As far as the many-universes interpretations are concerned, I would have felt that, again, I'm not satisfied with their avoidance of the various paradoxes, the EPR paradox and the Schrödinger cat paradox. And as far as interpretations which are dealing essentially with an idea of hidden variables, or uncontrollable variables, I would say that they could not even get as far as quantum field theory.

But in defence of the many-universes theory, it would be claimed, I think, by their proponents, that the paradoxes such as that of Schrödinger's cat are easily resolved because if you ask in any individual case whether the cat is alive or dead, the answer is both. And in one universe the cat is alive and in the other universe it's dead, and that seems to be a perfectly satisfactory explanation. In the ensemble interpretation the answer is . . . well, we can't answer.

I'm afraid I don't see it as satisfactory. I really must confess that I find the many-universes interpretation as bizarre. No, I'm sorry, I'm a hard-nosed physicist. Since one has no idea of what goes on in the other universes, they shouldn't be brought in.

It does of course have the other advantage that it may be able to make sense of the notion of the quantum mechanics of the entire universe – quantum cosmology. Now, in the ensemble interpretation, doesn't that present you with a difficulty, because we only have one universe, so how can we ever talk about the quantum mechanics of the whole universe?

Well, I think that is a problem, but it's one that we can face if we have a universe that goes on for an infinite extent (i.e. is spatially infinite). Because then we can only think of making local measurements. We can't ever, in a universe of infinite extent, expect to measure the whole of it. We make measurements in our laboratories over finite ranges. I think it would really be too much to expect that we can have a wave function to describe an ensemble of infinite universes. It would be beyond our comprehension.

Quantum cosmology really is a non-starter then?

Well, no I'm not saying that, because we can have a wave function to describe the whole universe, but we can only measure bits of it, and so our ensemble interpretation can still work. Provided we have a universe which is infinite in extent. If it is finite in size, then we may have problems, in that we could then think of a laboratory which would cover the whole of the universe. So, indeed, if we find, by observing the deceleration of distant galaxies, that the universe is in fact going to collapse again (and hence be finite in size), then we may be in trouble as far as our quantum mechanical interpretation of an ensemble nature is concerned. The difficulty with the many-universes interpretation, though, is that one is bringing in so many additional things that we can never find out about. You can never work in these other universes.

Of course, once again, the proponents would argue that, although the physical conglomeration of universes might appear to be a rather bulky and unwieldy structure, nevertheless the epistemology of the theory is extremely elegant and slim because we don't need to make many assumptions.

But the assumptions are so bizarre that I would say it is not at all slim, and I would also reiterate that unless you can actually observe anything in these other universes they should not be introduced. You see, in the ensemble interpretation one is saying that there's only a limited amount of information one can obtain. But in the many-universes interpretation one is saying that there is a plethora of information which one cannot obtain. And that's because most of it is in other universes – in fact, an infinite amount of it is in other universes.

So really you're saying that both interpretations forsake information; in the case of the ensemble interpretation we simply say that we can't answer questions about individual systems, in the case of the many universes one we can't answer questions about the other universes?

That's right, yes. I would choose having less information than having information that I can never find out about. But then I wouldn't even call it information, I would call it hallucination!

8

David Bohm

Before his retirement David Bohm was Professor of Theoretical Physics at Birkbeck College London and for 30 years has been an acknowledged world authority on quantum mechanics. He was responsible for recasting the EPR experiment in its modern form. Throughout his career Bohm has been a leading advocate of the hidden variables school of thought, and has written many papers attempting to formulate a detailed theory. More recently, along with his co-worker Basil Hiley, he has constructed a non-local theory of quantum mechanics based on the idea of 'quantum potential'. Bohm is also well known for his philosophical deliberations on modern physics.

Can you explain how your interpretation differs from, let's say, Bohr's Copenhagen interpretation of quantum mechanics, which I think we can call the official view?

Yes. Well, there really is no very clear official view. I would say that there are several variations. But the general idea is that quantum mechanics cannot describe 'actuality' – that is, what actually happens as a self-referent process. You see, if we say something 'actually happens', quantum mechanics can only describe what can be observed in a measuring apparatus.

Isn't that all that one would need from a theory – what we can observe or measure?

Well, yes; if you presuppose that's all you need. But there is a difficulty with this view. The Copenhagen interpretation only gives a formula describing the probability of what can be observed in a piece of apparatus. Yet the apparatus itself is supposed to be made of the very same sort of things we're studying (i.e. particles subject to quantum effects).

Atoms?

Yes, atoms. Therefore if you want to discuss the existence of the apparatus, you should in principle use another piece of apparatus to look at it, and so on and so on.

This is the famous infinite regress?

Yes. Now Wigner has ended that regress by saying that only when somebody becomes conscious of a phenomenon is it really 'actual'.

How do you feel about that particular interpretation?

Well, it is one way of looking at things. My own feeling is that there is an area where it is true, especially in human relationships; people becoming conscious of each other can have a tremendous effect on each other. But I don't think it's really true for the experimental situations on which physicists work in the laboratory. It does seem to me that at this level the universe is independently actual, and that we are part of it.

Do you think that the external world exists in some sense independently of our existence, and independently of our observations?

Every physicist really believes that. For example, he talks about the universe having evolved before there was anybody around to look at it, except possibly God. Now unless you want to attribute it to God, as Bishop Berkeley did (and most physicists don't want to do that), you're unable to solve the problem of how the universe exists without physicists to look at it – or without somebody else to look at it.

As I understand it, the dispute between Einstein and Bohr was that Einstein insisted that our observations merely uncover the reality which already exists, whereas Bohr says that our observations actually create that reality. So you're closer to Einstein's position?

Well, Bohr doesn't even say that. He says that we deal with
nothing but phenomena, appearances and regularities in
phenomena. And he says, essentially, that reality is ultimately
ambiguous and unspecifiable.

*But you would find yourself much more in sympathy with Einstein's
point of view, that our observations uncover a reality which in some
sense already exists?*

Well, I've already put myself in between Einstein and Bohr. I say
there is an area where our observations do create the reality, as in
human relationships: when people become aware of each other
and communicate they create the reality of society. But I think
that the universe as a whole does not depend on us to do that.

*It seems to me that by adopting this position, you abolish mind
altogether from the universe.*

No, I say mind is real, mind may be very real. I specifically said
that between people mind has a tremendous effect. It affects the
body, it affects human relationships, it affects society.

But it doesn't affect atoms?

I don't think it has a significant effect on atoms. At least the
human mind doesn't. Perhaps you could take the view, as Bishop
Berkeley did, that the mind of God was responsible for the
creation of all matter. But then we must not equate ourselves with
God!

When you talk in your book Wholeness and the Implicate Order,
*this wholeness is referring to both mind and matter – the matter which
exists around us. Could you say how mind and matter fit together in
this view of wholeness?*

Yes, you're referring to the implicate order. Perhaps we could
talk first about Descartes, who made a distinction between mind
and matter. He said that there was thinking substance which we
call mind, and extended substance which we call matter. Now
they are so different it's hard to see how they could be related.
You see, our thoughts have no extension.

Yes, we can't find where they're located in space, for example.

That's right. So what Descartes proposed was that God put clear and distinct thoughts into the mind of man. God was able to do it because he created both mind and matter – man and everything else – and therefore he could put these thoughts into the minds of men so that they could understand extended substance. Now when the notion that God may be used as an explanation of things was dropped, then there was nothing left. Mind and matter were left totally unrelated. However, the implicate order – the enfolded order – shows that mind and matter may nevertheless be looked at in a similar way. Quantum mechanics may see mind and matter as enfolded.

Could I ask you to explain what you mean by implicate or enfolded order? Can you give a simple example?

Yes. Well, the simplest example is that if you fold a piece of paper and make a pattern on it, and then unfold it you get all sorts of new patterns. While the paper was folded the pattern was implicit – in fact the word implicit means enfolded in Latin – and therefore we could say the pattern was enfolded. Now quantum mechanics suggest that this is the way that phenomenal reality comes about from a deeper order in which it is enfolded. Reality unfolds to produce the visible order and folds back in. It is constantly unfolding and enfolding at such a rate that it apears to be steady. And now you can say that I'm proposing that thoughts and feelings and mind work in a similar way. The very fact that we say a thought is implicit means it contains another enfolded thought. Right?

Yes, but enfolded in what? In what are our thoughts enfolded?

I'm going to avoid that question for the time being, and say I want to show a similarity of form between mind and matter. Now this is what Descartes did not have. His belief was akin to saying that thought is enfolded, and that matter is extended. However, I'm saying both are enfolded and both unfold, therefore they are similar in their basic structure, though they may be very different in many other ways. Their similarity in basic structure is what enables us to understand the possibility of their being related.

What you're saying sounds to me very much like oriental philosophy. Perhaps some students of Zen would find these ideas very familiar. Do you see your thinking in this subject area as giving support to oriental mysticism?

Well it might do so, yes. But I think that this idea of enfoldment has also been present in the West. You see, Nicholas of Cusa proposed a similar idea several centuries ago. He had three words: *implicatio* (enfolded), *explicatio* (unfolded) and *complicatio* (all folded together). And he was saying that reality has this enfolded structure: that eternity both enfolds and unfolds time. Now, I don't think that we should categorize things as East or West, but look at the ideas on their own merit. And I think that quantum mechanics specifically is suggesting this enfolded order. If you go into it as I have done, you can begin to make sense of some of the strange properties of quantum mechanics by looking at it this way.

Can you say why? What is the crucial feature of quantum mechanics that leads you to believe in the idea of enfolded order?

Well, it's the wave–particle duality: you may say that something can unfold either as a wave-like or a particle-like entity. The mathematics of quantum mechanics – if you look at it carefully – corresponds to this enfoldment. It's very similar to the mathematics of the hologram, you see.

I was going to suggest that the hologram seems like a very good example of implicate or enfolded order.

Yes, it's one of the best ones we have where we see that a pattern is enfolded into the photographic plate, and when you shine light on it it's unfolded into a visible image. Each part of the photographic plate contains information about the whole. So the whole is unfolding from each region.

So your view of the atomic world is that in some way all of the information about a particular physical system is encoded somewhere, but it's encoded in an obscure way that we don't normally have access to?

Yes, but by definition it must be obscure when we look at it in the usual way, because I think all encoding, such as that found in DNA for example, is very obscure when looked at on the large scale.

If we take the famous case of the position and the momentum of a particle: according to the Heisenberg uncertainty principle, we can choose either one or the other to be well defined, but not both at the same time.

That's right, we can encode these properties so as to let one or the other develop.

But you're saying that in reality both of these quantities have well-defined meanings, well-defined values, but somehow we can only get at one or the other in experiments?

No, not exactly. You see, another example of the unfolded order is a seed. If you take a seed it contains encoded information and what happens is that when it's put in the ground the substance of a plant develops from the air, the water, the soil, and the energy from the sun. These materials are just moving in their usual way, but with this tiny seed of information they start to make a tree instead of whatever they would have done otherwise. And now that tree can produce a seed which makes another tree, and so on. Now, you can't say definitely that the tree was in the seed: the kind of tree that developed – its shape and size – depended not only on the seed, but also on the whole environment. And now if you go to a forest you can see trees are continually growing, dying, and being replaced by new trees and, if you visit your forest every hundred years, you would say that apparently trees had moved from one place to another. In fact, they're continually unfolding and enfolding, and that is the picture I want to give of the motion of matter at the most fundamental level. I want to say that life, mind and inanimate matter all have a similar structure.

Now, as far as I'm aware there's no known experiment which cannot be satisfactorily explained by quantum mechanics as it is. Do you disagree with that?

No, but that's begging the question. If the only purpose of physics is to explain experiments, then I think physics would have been a great deal less interesting than it has been. I mean, why do you want to explain experiments? Do you enjoy doing it or what?

Well, I think my position, if I can be so bold as to put it, is that physics concerns making models, that we make models of the world about us, to help us to relate one type of observation to another. And we either have good models or less good models. And that there is no such thing as a 'real world' in the sense of something which exists 'out there' to which our models are mere approximations. As all we can ever do is to make observations, what more can we ever want of physics?

I think that observations and experiments are guided by our way of thinking, and the questions we ask are determined by our way of thinking. And for thousands of years people haven't asked themselves the right questions. In quantum theory we're now asking a certain kind of question and we're getting a certain kind of answer. We may be putting ourselves into a trap, you see, by restricting ourselves to this way of thinking.

So you think that by adopting a new way of thinking, a new approach to the subject of microscopic physics, we could perhaps construct a very different set of questions to ask, and maybe end up with a very different theory?

Oh yes. And that's happened before many times. If you go back to the motion of the planets – the old idea of epicycles led people to ask certain questions, and then Newton's laws led us to ask a very different set of questions. Statistical mechanics leads to one set of questions; and quantum mechanics makes another set of questions and so on. The questions you ask are determined mainly by the theory, by theoretical conceptions.

But usually a particular way of approaching a topic is followed until some experiment comes along that doesn't fit in with it.

I think that's presupposing that that's the only way. You may have to be banged on the head for 200 or 300 years before you'll change your ideas. For example, I think non-locality was obvious 50 years ago, but now only a very few physicists realize it's there. If they'd get banged on the head for another 50 years maybe more will realize it's there.

Let's talk a little more about non-locality. I wanted to ask you about your response to the Aspect experiment, which has been performed recently. Now, as I understand it, given Aspect's results we have to relinquish either what we might call objective reality – the external world existing independently of our observations – or locality – the idea that different regions of the universe can't signal to each other faster than light, crudely speaking. Which of these two are you prepared to relinquish?

I would be quite ready to relinquish locality; I think it's an arbitary assumption. I mean in the last few hundred years it has been given tremendous weight. If you went back 1000 or 2000 years, almost everybody was thinking non-locally.

But now don't we run into horrendous paradoxes such as being able to signal our own past?

No, that's only if we assume that the present theories are the last word. That's the whole point of considering new ways of looking at things: put your questions in a different form, and you won't get into these paradoxes.

So you want to abandon the special theory of relativity?

I don't say abandon relativity theory. I'm saying it's going to be an approximation to a much broader point of view, just as Newtonian mechanics is an approximation to relativity.

But you must certainly entertain the idea of faster-than-light signalling.

Yes. I have a view which would entertain that and yet not contradict any experiments which have been performed.

Can you think of any new experiments which could test this non-local feature of your theory?

That's a bit premature, because we're in a peculiar situation, as when Democritus proposed the atomic hypothesis several thousand years ago. If you had said then that we will not think about this unless we can propose an experiment to verify the hypothesis then that would have been the end of that idea. There was no way at that time to propose an experiment. Even if anybody had been ingenious enough to propose one, there was no equipment available that would have made it possible anyway. And yet the idea was still valuable.

So are you saying that, in fact, not only are we not capable of testing this faster-than-light signalling, but that you can't off-hand think of a way in which it could be done?

I think you must entertain a notion for a long time before you can do something new. If you say, 'I will only think of something the minute you propose an experiment – otherwise I won't think of it', how will you ever propose anything new? It often takes many, many years to be able to see what sort of experiments could be done. It took 2000 years to get enough content into the atomic theory to propose an experiment. So what would you have said? That nobody should think about it until suddenly an idea would occur for an experiment? Experiments would never have been done if nobody had thought about it.

But do you believe that by using quantum effects and bringing about faster-than-light contact between separated systems, it will ever be possible to literally send signals into the past?

No, I think that those paradoxes will not arise in the way I have formulated the question. Those causal paradoxes only arise if you say that relativity is the absolute truth.

How precisely could this faster-than-light signalling come about?

Well, you see this would require some historical explanation. I proposed another interpretation of quantum mechanics, an alternative interpretation, in 1951. There are two stages to this: first, as applied to particles and, then, as applied to fields. Now, in the first stage, I said that an electron is essentially a particle, but, in addition to all the other potentials it has, such as the electro-magnetic potential, it has a new kind of potential, which at that time I called the quantum potential.

Which, crudely speaking, we can think of as being something that would jiggle the electron around?

Yes. Now the quantum potential has new properties, and the first of these is that its effects do not depend on its magnitude but only on its form, so that it may have big effects at long distances. In this way you can explain, say, the two-slit experiment.

This is normally explained, of course, by proposing the interference between waves passing through the two slits.

It's not explained, it is merely described. If you said it was a wave, that would be an explanation. But since the electrons arrive as particles, it is no explanation. It is merely a sort of a metaphorical way of talking. Right? There is no explanation. We should say that quantum mechanics does not explain anything; it merely gives a formula for certain results. And I'm trying to give an explanation.

How does the quantum potential explain the interference?

Well, the quantum potential, which is carried as a wave, can affect particles even quite far away from the slits, because, as I've said, its influence depends on the form not the magnitude. Now, the quantum potential or wave is quite different if the second slit is open than if it's closed. So particles which pass through can be deviated by the quantum potential even a long distance from the slit, in such a way that we produce these interference patterns. Now this shows a new property of wholeness that in some ways I agree is similar to what Bohr said, but I'm proposing to give an explanation of it.

So part of the information carried in this wave or potential would be the experimental arrangement?

The experimental arrangement, yes. Also the states of all the other particles in the system, and so on. So therefore you have

what I call a non-local connection. This information brings about
the new quality of wholeness, in the sense that each part now
moves in a way which reflects the state of the whole. It may be
that the connection is very weak under ordinary circumstances,
but special conditions exist in which it can become quite strong,
such as superconductivity or the two-slit experiment that I just
described.

*This wave that you introduced many years ago is clearly not the same
as the wave which we're familiar with when we talk about the wave
aspects of matter?*

No, it's a new kind of wave which I call 'active information'. The
notion of active information is already familiar to us from computers. Also, if I tell you something and you do something, that's
obviously active information. If I shouted 'fire', everybody
would move, so we know that in living intelligent systems, and
in computers, active information is a useful concept. Now what I
am proposing is that matter in general is not so different.

*We're familiar with other types of potentials like electric potentials,
and gravitational potentials, how does your quantum potential compare with those?*

Well, you see it's similar in that it obeys certain equations,
though more subtle. It's different in that it does not necessarily
fall off with the distance, and that its effect is active and does not
depend on the intensity of the potential but only on the form.

So there really is nothing else like this in physics?

Yes, but we've often been in that situation, where something has
been introduced that wasn't there before.

*Earlier you implied that although the quantum potential idea enter-
tains the notion of faster-than-light signalling, it would not conflict
with our present experimental results. Can you tell us how this is
possible?*

Yes. Well, this involves an extension of the idea of quantum
potential to a field, the entire field of the universe, which I call the

super quantum potential. And this would take some explanation. But, basically it will bring about a connection of fields at different points instantaneously. Now it does not violate the principle of relativity in any experiment, because one can show that the statistics of experiments as done so far in the present system of quantum mechanics will still come out in agreement with the theory of relativity.

That is, forbidding signalling faster than light?

There's no way to signal because we are using only statistical experiments anyway.

We have no control over the influences that propagate faster than light?

Yes that's right. So long as the present type of experiment is done, the theory of relativity will still be saved. But if we could manage to get deeper than that then we might find that there was something faster than light. You see, we would then say that relativity and quantum mechanics have the same limit, namely the limit of statistics.

The usual objection about faster-than-light signalling is that, if we can encode and transmit information, then that would lead to paradoxes; whereas you're saying that basically we don't have that control over the microworld and everything is fuzzed out by the unpredictable nature of quantum phenomena?

Yes, and one can even demonstrate that therefore there is no way to get any inconsistency, and that, if we had some hold on the thing that was deeper, we could then get beyond these limits.

It seems a little ironical that you are, if not contradicting Einstein's special theory, at least drastically modifying it, perhaps against the spirit of the original theory. What do you think Einstein would make of this?

Yes, well I don't think anybody can necessarily expect everything to turn out the way he expects it. Quite a few things turned out that way for Einstein, but he can't have everything right!

One argument against the use of your quantum potential is that it sounds like a very complicated thing: it doesn't have a simple set of equations in the same way as, say, an electric field does.

The equations are just Schrödinger's equation either for the one or the many-body problem. Nature is telling us that the simple idea of the electric field is too simple! And the point I'm trying to make is that nature has a complexity and subtlety which approaches that of the mind. I'm trying to say we have had too simple a view about nature.

You think that this is perhaps due to the Newtonian tradition of reductionism, of chopping the world up into lots of little pieces?

That's right. I don't know whether Newton was behind it, but those who followed him certainly did that.

Whereas you would feel more in sympathy with, let's say, a synthetic or holistic view, where we have to take into account the total system to understand any component of it?

That's right. Yes I'm glad you brought that up, because now we have to ask, 'how do we explain the ability to analyse the world into independent parts in ordinary mechanics?' The answer is that when the wave function has a certain property which I call factorizability – that's a mathematical term – you find that the various parts behave independently. Now, under ordinary circumstances that's a good approximation, but quantum mechanical experiments are so designed as to produce situations where the wave function is not factorizable, so they can demonstrate wholeness.

Could I come back to the Aspect experiment? Are you saying that, when the two photons travel in opposite directions and reach relatively widely separated points, their cooperation can be attributed to a signal passing between them faster than light?

Well, I think the word signal is wrong, because it has a certain connotation which means that you can transmit messages. This would not be that definite, but there would be a connection, I prefer to use the word connection. You see, a connection is established such that what happens to one particle will affect what happens to the other one. Now, conventional quantum mechanics doesn't explain the Aspect experiment. It merely gives you a system of calculating (the results of that experiment). You see, I think we should distinguish between explanation and systems of calculus, and quantum mechanics is a calculus that enables you to predict statistical results. But it has no explanation, and Bohr emphasized that there was no explanation of any kind.

But is there ever explanation in physics? I mean, don't we simply make models and invent language for them?

But models explain the thing in the sense that they show how it comes about; the explanation makes it intelligible. Quantum mechanics says that nature is unintelligible except as a calculus, that all you can do is to compute with the equations and operate your apparatus and compare the results.

Can you think of another area of physics, say a simple area, where you think that we actually have an explanation?

Well, a lot of classical physics gives an explanation in so far as it's correct.

In what way though? Isn't it just language and models relating observations? Where is the real explanation? We use the word 'explanation', but it seems to me rather meaningless, and that all you're really doing is relating observations together successfully.

I don't think so. You see, I think observations are a secondary affair. I can't understand the tremendous emphasis in modern physics for putting observations first. I think it's the positivist philosophy which has done it. You must admit it mostly began this century. If you had gone back 200 or 300 years, everybody would have understood what explanation is and nobody would have understood what the positivists were trying to do.

It's true. But supposing we take a particular example – why does the apple fall – and we say the explanation is because there's a gravitational field, and the Earth is acting on the apple, then we're still left with the problem of explaining the gravitational field.

Yes, but we are at least giving an account of what actually happens: we say there is an apple, and it follows a path, and we understand how the apple gets from here to there by passing through stages in between. Now if we were to take quantum mechanics, we would say that explanation has gone, we have an apple here, we have another apple on the ground, we have no notion of how one would connect up with the other, we don't even know whether it's going to happen, but we have a calculus which gives the statistics of the number of apples arriving in certain places. Now this is similar to the insurance company saying we have statistics on how many people of a certain category will die in a certain year, and that's all we care about! But that is not an explanation.

But if we go back to the apple and we think of this purely classically, after all we can only make observations on the apple, and we can make measurements of where it is at certain times and so on: at the end of the day, if we have a successful theory, then it will relate these observations together.

I think that's a secondary affair. It does that, but more importantly it gives a conception of what is happening.

Oh, it gives one. It gives us a simple mental image of what's going on, namely the apple is falling in a continuous trajectory to the ground. But isn't this image simply an illusion?

Well, what are the calculations then?

The calculations are a model that enable us to relate these observations together.

Why do you want to relate them?

Because it seems to me that physics is about making observations of the world.

Why is it about making observations? I mean, that's an idea which started a few hundred years ago. People hold on to it because they've been taught it by their teachers. But why do you say that?

Well, because for experimental physicists, it's their profession to make measurements of the world.

But physics did not begin purely with experiments, it began with people asking questions. I mean, there would have been no experiments if nobody had asked these questions. People were interested in the world from a much broader point of view.

This raises Popper's idea about what we can regard as scientific. He argues that you have to be able to show the theory is potentially falsifiable, and that depends on being able to make observations which could contradict the theory.

That's Popper's idea. I'm trying to say, why should we take him as the authority? There are all sorts of ideas which people have had, and Popper has proposed an idea which has some merit, but it needn't be the absolute truth. If one says that Popper has given the absolute last word as to what science is, then why should I accept that?

To summarize, then, I think that in the absence of any experiment to the contrary, all we are debating here are really different philosophical standpoints?

Yes, well, originally the word philosophy meant love of wisdom. Now it becomes a sort of technique. Also I think our modern age is falling into reducing everything to techniques, and it takes away the significance of everything. I think that people have gradually fallen into that, and have said that anything else which doesn't fit that simply is of no consequence. You must notice this has developed historically. You can't regard it as an absolute truth.

But although we sit here discussing what we might call philosophy – and there's been a great deal of discussion about the conceptual foundations of quantum mechanics which seems to me to be mere philosophy – nevertheless if I'm correct, you do foresee a time in the future (we don't know how long in the future) where real experiments will be done which will actually expose the weaknesses of the present interpretation of quantum mechanics.

Yes, but I think that any fundamental new experiments arise from philosophical questions. If you go into history, in the Greek times, science was largely speculative. People then corrected that by bringing in experiments. Now we've gone the other way and said experiments are almost the only thing there is to it. So, in effect, we have gone to the opposite extreme. Science surely involves several things? It involves insight into ideas, and this insight precedes experiment. If you exclude philosophy you will eventually exclude these things too. The only insight available now is through mathematics: that's the only place people allow themselves any freedom. They can play around with mathematics as much as they like without experiments. I saw an article in the *New York Times* a few months ago where they said we have supergravity, and they said it looks promising, but that we won't be able to say anything definite for 20 years! So nobody minds, as long as it's mathematics. People believe that mathematics is truth, but anything else is not.

Well, it's true that mathematical elegance is a criterion which people have used in support of a theory where experiment is lacking.

But if you will allow mathematical elegance, will you not allow elegance in the conception? Every physicist has at least a tacit philosophy but the present generally accepted philosophy is extremely inelegant. It's really crude.

But nevertheless – and I'm sorry to keep coming back to this – do you feel that there will be a stage in the future where it will be possible to do experiments to discriminate between these different interpretations?

I think there will be, but there won't be if you don't first consider these ideas seriously in the absence of experiments.

But you don't have any particular experiments in mind to propose at this stage?

No, but I'm trying to say that if everybody took that attitude, saying we will not listen to anything anybody has to say until he proposes an experiment, then nobody can ever propose anything fundamentally new.

9

Basil Hiley

Basil Hiley is Reader in Physics at Birkbeck College, University of London. His research interests are in solid state, liquid state and polymer physics, as well as the conceptual basis of quantum mechanics. A long-standing collaborator of David Bohm, he has for many years rejected the conventional interpretation of quantum mechanics and attempted to construct a theory more in keeping with 'common sense' realism. His recent work with Bohm on the non-local 'quantum potential' is a direct challenge to the orthodox view.

Aspect's recent experiment suggests that the traditional approach to quantum mechanics is alive and well, and that we can continue to use it with confidence. Now, in your quantum potential theory you seem to take a radically different approach. Why is it that you distrust the conventional interpretation of quantum mechanics?

I think distrust is the wrong word. If anybody came to me and said I want to solve a certain physical problem, I would recommend that they go ahead with the conventional interpretation because we know it works and gives the correct answers. But when you look at the conventional interpretation, and you try to undertand what is going on when electrons produce an interference pattern for example, you have no physical way of explaining the formation of this pattern.

Why do you feel it is necessary to say what the electron is doing? After all, in physics, and not just in quantum mechanics, the only access we have to the world is through our instruments and experiments, and the only data that we have to go on are our experimental results. Why do you want to push the model of the external world so far that we need to talk about what the electron is doing even though we can't actually observe this? Aren't our observations enough?

No. I think that what we have to try and do is to build up a model with which we can reinforce our intuitive ideas about the physical world. I've been brought up as a physicist, and I feel that intuition has always helped a great deal. When I look at quantum mechanics I find it completely counterintuitive. We just have a prescription – a set of rules: we have a wave function which is supposed to describe the state of the system; we have an operator which we then apply to the wave function; and we get certain predicted experimental values out. But this doesn't help me in understanding the two-slit experiment, for example. What exactly is the electron doing when it passes through the slits? Does it go through one slit, or does it go through both? Now, these are questions which are extremely important in trying to get a feel of what is actually happening.

Let's try to make this absolutely clear. In the conventional, or Copenhagen, interpretation one can talk about either the position of an electron, say, or the momentum of an electron, but not both simultaneously. And that's because we don't know where the electron is or how it's moving; it's meaningless to even talk about the electron having a well-defined position and a well-defined motion simultaneously. Now you're saying that the electron really does have a well-defined position and motion even though in practice we can't determine both of these quantities at the same time. Is that right?

Yes. The model I have been looking at was originated by de Broglie and subsequently developed by David Bohm. The difficulty with the usual approach is that we can only talk about 'observations' or 'measurements' and cannot say what happens in between. I feel we need to explore ontologies in which we can raise such questions and this could mean that we can attribute a precise position and momentum to a particle even though for the observer these are unknown.

This is the so-called quantum potential idea? Can you summarize the essential features of this approach?

First of all we imagine that there is an actual particle that has both a definite momentum and a definite position. We then take its wave function, and, rather than regarding this as a means of calculating probabilities, we treat it as a real field, something analogous to the electromagnetic field. The field can then influence the behaviour of this or other particles. Technically this is achieved by an equation of motion derived from Schrödinger's equation. It contains an additional potential which we call the quantum potential, since it modifies the classical behaviour of the particles to produce results consistent with quantum mechanics.

What sort of wave or field is this?

Although I have used the analogy of the electromagnetic field, it actually has properties which are very different from the electromagnetic field.

What properties are these?

Perhaps I can best illustrate them through an example. We know that if we pass electrons through a screen with two adjacent slits, the result on the other side looks very much as if we're getting waves interfering with each other. And, indeed, the orthodox theory actually uses the wave function to describe this particular wave phenomenon. But what we actually see at the other end is the arrival of individual electrons. So the wave is really an *average* of how a beam of individual electrons behave, and the intensity of the wave corresponds to the number of electrons arriving at that particular spot in a given time interval.

Now, orthodox theory says that you cannot actually predict how each electron will arrive at the screen. But what the quantum potential does is to enable you to calculate the set of individual trajectories that gives rise to the interference pattern. You can therefore look at the form of the quantum potential from the calculations that you use. The quantum potential will contain things like the slit width, the distance between the slits, and the momentum of the particle; in other words it appears to have some information about the environment of the particle. It is for this reason that one tends to regard the quantum potential as arising from a field that is more like a field of information than a physical field.

Perhaps I can take this analogy a little bit further. Suppose we imagine that we have a ship which is guided by radar waves; the radar waves are fed into the ship's computer, and the ship then adjusts its direction depending upon the information that it receives from the radar waves. Now, we're trying to suggest that the quantum potential arises from waves that are more like radar waves. The quantum potential carries information about the environment which is fed into the electron so that the electron then adjusts its movement in order to produce the bunching effect we observe on the screen.

So the motion of the electron isn't forced upon it by the quantum potential. The potential just carries the information to tell the electron how to move?

Yes, it's an information potential. The more traditional way in physics is to think that the electron is pushed around by the field, just as water waves can push a ship around. The quantum potential doesn't work like that, because you can actually multiply the field by a constant and it doesn't change the force on the particle. So it's not an ordinary classical force pushing the electron around.

This quantum potential seems to be quite unlike anything we have encountered before in physics. Indeed, it seems to be rather remarkable. If we think of the electron as being like the ship, moving under the guidance of the information carried by this potential, it rather makes the electron seem a bit like a super-computer. Can we really imagine that such a simple thing as an electron, which is supposed to be structureless and have no internal parts, can respond in such a complicated way?

When I first began to think about this idea, I remember that Richard Feynman had already pre-empted us in saying that he thought of a point in spacetime as being like a computer with an input and output connecting neighbouring points. The point would have a memory for all the fields and particles that are possible and would actually act like a computer. So he has each point in spacetime acting like a computer! I am only suggesting that the *electron* may act like a computer!

Of course present experiments fail to reveal any structure within the electron down to a distance of about 10^{-16} centimetres. But remember we've still got to go down to gravitational lengths which are about 10^{-33} centimetres, so there's still room for a lot of structure, even though it is going to be rather small on our scale.

So, you think that a particle like an electron could actually be a composite body with internal parts that can act rather like the components of a computer?

I wouldn't like to push the analogy too far, but it is a possibility.

I have a rather naive question now. I think it's very nice this analogy of the ship with the radar, but of course to enable a ship to respond to radar signals it still has to have some motive power of its own. So if the electron gets the message from this quantum potential which says: 'Move to the left!', well, how does it move? What is its motive power?

The motive power comes from the quantum potential itself.

But I thought the quantum potential just triggered a response in the electron, not drove the electron?

I have not made myself clear. It is the wave field that triggers a response in the electron. This is translated into a quantum potential which is part of an equation of motion. In terms of this equation the quantum potential does give rise to a driving force with energy coming from the self-activity of the electron. But I don't like pushing too far down this line, because I have a slightly different image of the electron. I don't think the electron can be completely separated from its environment. You see, one of the things about quantum theory that Bohr emphasized is that we have to look at the *whole* experimental situation. And what seems to be coming across from the quantum potential approach is that we can actually take his idea a bit further. If we cannot separate particles and treat them as independent entities, we have to regard them as aspects of the total situation. It's the whole system that responds, so we shouldn't think of the electron as having something which drives it from within. That would be like going back to a mechanistic view of cogwheels or computer parts inside the electron.

There was a time when it was proposed that maybe the quantum uncertainty of an electron was due to it being jiggled about by random forces in its environment, in the conventional way that a wave may toss a cork around on the surface of the sea. If we think of the electron as following a zig-zag path, then it's easy to see if it's subject to random forces, it can be forced along a zig-zag path. But you seem to be saying that the quantum potential tells the electron how it's got to zig-zag about, but we can't find any motive power to cause the zig-zagging.

We always have the zero point energy. We know the vacuum state is actually full of energy, and the orthodox theory exploits that energy.

Yes. That's hard to push through in detail though, isn't it? For example, you would expect there to be a difference between neutrons and protons, and yet their quantum mechanical behaviour is very similar.

But I'm not thinking of this in terms of the electromagnetic background, because the quantum potential arises from a field that is not like an electromagnetic field. It seems to be very different; it seems to be much subtler than that.

So this zero point background that you're talking about is some sort of background of the quantum potential field rather than the zero point energy associated with other types of more familiar fields like electro-magnetic fields?

That's right.

If we turn directly to Aspect's experiment, in which one is dealing with a two-particle system rather than a one-particle system, the experiment shows that we have to make a choice: we can either reject what we might call 'reality', the idea of the external world existing independently of our measurements, or we can reject locality, the idea of all signals and influences travelling no faster than the speed of light. Now, the quantum potential idea as I understand it, attempts to retain at least a vestige of the old idea of objective reality, but the price one has to pay is that you end up with a measure of non-locality. Is that right?

Are you suggesting that quantum mechanics does not have that non-locality in it?

No, I realize that quantum mechanics has an element of non-locality as well, but of course in the Copenhagen interpretation one is usually quite happy to abandon the naive version of reality. And so it's possible to make the Aspect experiment consistent with the absence of faster-than-light signalling.

If you're essentially saying that we can calculate the probabilities using quantum calculus then I agree with you wholeheartedly. We can do that. What is not clear to me from the orthodox theory is how to understand Aspect's distant correlations. What the quantum potential does, is to show unambiguously that there is a non-local connection between the two. I know that if you go back to Einstein's view, namely that reality is a description in spacetime with only local interactions, then this would rule out the quantum potential point of view. That, incidentally, is one of the reasons why Einstein didn't think too much of the quantum potential point of view.

Does that worry you?

No, it doesn't. We now have experimental evidence to show that reality does have some form of non-local element in it. What we have to do is to ask why most experiments only reveal local connections. We have already begun to see how that could be explained by extending the idea of the quantum potential to the quantum theory of fields.

Supposing one could push this programme through – and it is of course tentative at this stage – but supposing we could push it through, it does seem to lead to the possibility of faster-than-light communication. If we accept the theory of relativity this could enable us to communicate backward in time. Now this seems to be a recipe for all sorts of causal paradoxes, and it seems a high price to pay for hanging on to some vestige of naive reality.

The quantum potential will not have any causal paradoxes in it because it essentially requires an absolute spacetime in the background, a quantum aether of the type suggested by Dirac. Let me explain. We take the field theory and construct a superpotential from the fields. You can then show that the superpotential (which is governed by a Schrödinger superwave equation) is in instantaneous contact with all particles (i.e. non-local). But when you work out the statistical results of typical quantum experiments you find that they are still Lorentz invariant (that is, they obey the theory of relativity). So in other words, relativity in the quantum potential approach comes out as a statistical effect, not as an absolute effect.

So there is no way in practice of being able to send signals faster than light?

That is not clear. There is no way that we can see at the moment. But if you have an absolute spacetime, or an absolute time, in the background then you don't get into causal loops. So the causal paradoxes won't arise in this theory. But you will have instantaneous connections, and the question is: what do those instantaneous connections mean? It is possible that we might be able to find other experiments that exhibit those instantaneous connections.

But if we take the behaviour of ordinary clocks as normally understood within the context of the theory of relativity, then instantaneous communication would amount in practice to communication backwards in time, would it not?

The point is that clocks are actually macroscopic collections of particles; they are statistical in the way they function, and they will not be able to detect these instantaneous connections.

No, a clock wouldn't, but is it not possible that one can devise a communication system which, although in your absolute spacetime would produce an instantaneous connection, nevertheless in the reference frames as normally used by clocks in special relativity this would amount to signalling backwards in time?

It's not clear to me that such a possibility exists. If we go back to Aspect's experiment, although the quantum potential shows there is an instantaneous connection, when we look at the statistical properties of the particles at each end of the connection, they (the particles) appear to be independent; it's only in the correlations that we see the non-locality. It's not clear to me that those correlations can ever be transformed into a signal which makes things go backwards in time.

At the moment, of course, it's not possible to use these correlations actually as a signalling device.

Correct.

And in the conventional interpretation of quantum mechanics that would never be the case. But it seems that with your interpretation it is in principle possible, although in practice you can't think of how to do it.

Well, I think this is of some merit for our theory, because it'll make us think very carefully about whether we can do this kind of thing or not.

It does seem as though you're aimed for a head-on clash with the theory of relativity.

I don't see it that way because, as I say, at the moment it looks as if it's the statistical effects which give us relativity. The problem is how are we going to design experiments which will go beyond this level to see these instantaneous connections. That's not clear at the moment. What is clear is that the quantum potential faithfully reproduces the results of quantum mechanics in our present experimental regime. It is not at this stage doing anything different.

So am I right that the only place where the results of quantum mechanics would differ from your theory is in the area of these instantaneous communications – in the very area that's going to get you into trouble with relativity?

The trouble is that in the orthodox interpretation of quantum theory we can't ask the question of what is happening between two separated systems. I cannot even think about the problem in the formalism as it is now, because I just have a wave function. And from that wave function I know how to work out the correlations, but I don't know what is going on underneath, so I can't raise the question. Now, maybe you think that we shouldn't raise the question, but if we've got a theory which produces exactly the same results as the orthodox theory, then it seems to me that we should explore this further and find out whether we will get any new physics. Maybe we won't and then you could argue that it's all been a waste of time. But at least we have a different view on this question.

OK. We'll let that point rest. But what advantages do you think your approach has other than providing us with a tidy model of reality?

What the orthodox approach has always left us with is this so-called measurement problem. And if you look back through the literature you'll find almost 300 papers on trying to solve the measurement problem. What's more, the exponents of the orthodox theory disagree over whether there even is a measurement problem or not.

This is where we bring the observer into quantum theory in an explicit way.

Yes. Now, when you're talking about the measurement problem, you have to remember that the orthodox theory says that the wave function describes the state of the system. And you then use your apparatus to determine how this state develops. When you use the apparatus you find that the state develops into what is called a linear superposition. Let me take the following situation: suppose you have an experiment which gives you two possibilities . . .

Suppose we take live cat/dead cat from the Schrödinger cat experiment?

. . . That would do, yes. You've got two possibilities: either the cat is alive or the cat is dead. If you now try to calculate what happens in the quantum mechanical formalism, you find that the state function for the cat at the end of the experiment is a linear superposition of a cat alive and a cat dead.

That means these two states are somehow overlapping each other.

These two states exist together in some way, yes. Now, when you open up the box containing the cat, you then see whether it's alive or dead, and that is referred to as 'the collapse of the wave function'. You cannot bring about that collapse of the wave function within the orthodox theory. So this has tempted people as distinguished as Wigner to suggest that perhaps (the act of) looking is a very important feature of quantum mechanics; namely, that somehow consciousness enters into the situation. When consciousness enters, the cat is either alive or dead, but before that it's in a sort of a state of suspended animation, being neither one thing nor the other.

I take it you don't like the idea of introducing mind into physics?

I don't see why mind should be introduced into physics at this level. Another idea people have is the many-universes interpretation of quantum theory. That is to say, when you look in the box what you discover is whether you're in one branch of the universe or another. One branch would correspond to the cat being alive, the other branch would correspond to it being dead.

The world is split into its two alternatives?

That's right, and we just happen to be following one of those alternatives. I'm not very keen on that idea because we seem to be producing many universes of which only one is observed by us. And so we have a rather strange situation. Now in terms of the quantum potential, we don't run into that difficulty. Because we have an actuality, namely the particle, and if the particle is in one of those waves, then as far as the information (quantum) potential is concerned no information feeds back from the other wave packet that we normally use in quantum mechanics (i.e. the part of the wave function corresponding to the other branch of the bifurcated universe).

Don't they interfere with each other?

They could eventually have the possibility of interfering with each other, but the point is that, when the particle is in one wave packet, as long as it is well separated from the other wave packet, they will not interfere. However, if those two wave packets are now allowed to overlap then of course there is the possibility of interaction between the two. But now when we make a measurement, one of the things that happens is that an irreversible process takes place; it is that irreversible process which is the key to the collapse of the wave function in the quantum potential approach. The 'empty' wave packet now cannot ever be brought back again to overlap the wave function of the one that had the particle in.

Why not? Because it's suddenly disappeared from the universe?

Perhaps we shouldn't talk about it actually disappearing from the universe. Rather the information in the 'empty' wave packet no longer has any effect, because during the act of measurement the irreversible process introduces a stochastic or random disturbance which destroys the information of quantum potential of the wave packet.

So it's not so much that a part of the wave disappears, it's that it gets scrambled up among other things in an irreversible way. The wave hasn't gone, it's just got completely interwoven with other waves, losing its original information.

I would accept that, yes. It's not active any more. And we've tried to introduce a distinction between active information and inactive information. That is, when an apparatus has undergone this irreversible change, one wave packet becomes inactive.

So instead of a portion of the wave disappearing (as a result of the act of measurement), it just becomes impotent?

Yes, that's right.

Let's look at the microscopic quantum scale again. You've said that a particle such as an electron does in fact have both a well-defined position and momentum. Yet we know from Heisenberg's uncertainty principle that both of these cannot be measured simultaneously. How do you explain that?

Well, that would just be a statistical effect. You see, when you bring the measuring apparatus into an experiment you have a many-body system. And the many-body system is necessarily thermodynamic in nature, so you can never hope to know where all the particles of the apparatus are. The very process of measurement or preparation of a system in a given momentum state, say, means you will then have all this uncertainty in it, and you can never be sure exactly where the particle is. We've always got an ambiguity because of this thermodynamic situation.

The uncertainty is introduced by the apparatus?

By the apparatus, yes.

It's our clumsiness in probing the system?

That's right. So in principle it would be causal in this interpretation. But in practice, because we are a thermodynamic system and our apparatus is a thermodynamic system, we cannot hope to determine the precise effect.

I can't then see how Planck's constant comes in because it seems that if quantum uncertainty is purely thermodynamical it's just a classical effect, and I can't see why there should be any preferred scale of action.

To me the value of Planck's constant is not really relevant to quantum mechanics. I know I'm committing a heresy here, because a lot of people are under the impression that if you put Planck's constant equal to zero you will just recapture classical mechanics from the quantum formalism, and nothing could be further from the truth.

But nevertheless it is a fundamental constant of nature which has a value, and if its value was different then the world would be a rather different place.

I agree, but the quantum potential does contain Planck's constant. And therefore if Planck's constant changed its value the quantum potential would change its value.

But the question we were dealing with a moment ago concerned Heisenberg's uncertainty principle in that if we carry out a measurement on a system and it's the clumsiness of the apparatus – in a classical thermodynamic sense – which introduces the apparent quantum uncertainty, why is it on a scale determined by Planck's constant? It seems a little bit mysterious why it should be that particular scale – if the uncertainty is purely a classical effect.

But at the moment we're essentially getting this quantum potential out of the Schrödinger equation, which already has Planck's constant in it.

Yes, but after all if we just go to a sort of classical interpretation of measurement, and we have a particle, and we're trying to measure its position and its momentum and so on, and we find that we're dong this in a rather clumsy way, then there's a degree of uncertainty in the results. And, of course, we know from thermodynamics that this is often the case. But if we imagine refining our apparatus more and more, and getting ever more precise results, quantum mechanics tells us that there is an irreducible uncertainty, and that's where Planck's constant enters. From what you say about the apparatus causing this disturbance, I can't see any reason for an irreducible level of uncertainty. Why should there be any particular scale of action?

This is a good question. I take your point. I think I agree with you that it can't be just the irreversibility. But remember we are using Schrödinger's equation to derive the quantum potential and since it contains Planck's constant our analysis will also contain it. So you are essentially asking me to explain why we need the Schrödinger equation. I don't know the answer to that one.

GLOSSARY

Action at a distance. The concept of two, separated systems exerting physical effects on each other. In modern physics direct action at a distance is replaced by field theory in which separated systems interact only by stimulating influences to propagate through a field which extends across the space between them. For example, the moon's motion, acting through the intermediary of its gravitational field, raises the ocean tides.

Aether. A hypothetical medium once thought to fill all space, thereby defining a universal frame of reference relative to which a material body's velocity through space could be defined. Electromagnetic waves were regarded as vibrations of the aether. The aether concept was rendered irrelevant by the special theory of relativity.

Aspect's experiment. An experiment performed in 1982 by Alain Aspect and co-workers to test the conceptual foundations of quantum mechanics by checking Bell's inequality for pairs of photons emitted simultaneously in single atomic transitions. (See p. 17 for full description.)

Bell's theorem (or inequality). After John Bell, who in 1965 proved certain very general restrictions, in the form of mathematical inequalities, of the degree to which the results of measurements performed simultaneously on separated physical systems can be correlated, given certain assumptions about the nature of physical action and the nature of reality.

Causality. The relationship between cause and effect. In classical physics, an effect is restricted merely to follow a cause. In relativistic physics causal connections are additionally limited by the finite speed of light. Events which cannot be connected by influences travelling at the speed of light or less are causally independent. One cannot affect the other.

CERN. Acronym for the Centre Européen pour la Recherche Nucléaire near Geneva in Switzerland, where some of the world's most powerful subatomic particle accelerators are situated.

Conservation of momentum. A fundamental law of both classical and quantum physics, which requires the total momentum of an isolated system to remain constant, whatever internal changes may occur in the system. In classical Newtonian mechanics, momentum is defined as mass × velocity.

Copenhagen interpretation. The interpretation of quantum mechanics associated with the name of Niels Bohr and his research school in Copenhagen during the 1930s. The Copenhagen interpretation is usually accepted to be the conventional viewpoint in spite of the continuing challenge to its position. (See p. 31 for full description.)

Einstein–Podolsky–Rosen experiment. A thought experiment devised by Einstein and colleagues in 1935 designed to expose the pecularities of quantum mechanics as interpreted by Bohr. The experiment, which consisted of measurements performed simultaneously on two quantum systems that at one time interacted and then moved far apart, forms the basis for Aspect's real experiment. (See p. 17 for full description.)

Electrodynamics. The theory that treats electromagnetic fields together with their sources – electric charges, currents, and magnets. Electrodynamics takes into account the motion of the sources, the propagation of the fields and the interaction between sources and fields.

EPR paradox (or experiment). See Einstein–Podolsky–Rosen experiment.

Faster-than-light signalling. Hypothetical mechanism involving physical effects propagating faster than light, thereby enabling events to be causally connected that would otherwise be regarded as physically independent according to the theory of relativity.

Heisenberg's uncertainty principle. After Werner Heisenberg, this is a mathematical formula that describes an irreducible level of uncertainty that is always present (for quantum reasons) in certain pairs of dynamical quantities when they are measured together, e.g. the position and momentum of a particle.

Infinite regress. Philosophically unpalatable outcome of an argument in which each step depends logically on a succeeding step, continuing in an unending sequence.

Irreversible process. In some physical systems, e.g. the swinging pendulum, the processes of interest can also occur in reverse. In others, e.g. the diffusion of two different gases into each other, the process is irreversible.

Locality. A physical restriction on the way in which events can causally influence each other. In a general context, locality refers to the idea that events can only influence other events in their immediate vicinity. It also has a more restrictive meaning. If all physical effects are assumed to propagate no faster than light, two spatially separated events that occur simultaneously cannot be causally connected. Hence an event can only be instantaneously connected to another if it is at the same spatial location.

Lorentz invariance. After H. A. Lorentz, this is a mathematical concept connected with the symmetry properties of theories. It relates the values of physical quantities observed in one frame of reference to those observed in another, in a way consistent with the principles of the special theory of relativity. A theory must possess Lorentz invariance if it is to comply with the special theory.

Non-locality. The hypothetical circumstances in which locality fails. Some quantum processes have non-local flavour in that spatially separated events can be correlated, but usually this is assumed not to violate the more restrictive definition of locality concerning instantaneous causal connection between spatially separated events.

Planck's constant. A universal constant of nature, denoted h, which quantifies the scale at which the quantum effects are important. It is present in all mathematical descriptions of quantum systems, and may appear in a variety of contexts, e.g. as the ratio of the energy of a photon to the frequency of the light wave.

Quantum field theory. The quantum theory applied to fields, such as the electromagnetic field. Quantum field theory forms the basis for current understanding of high-energy particle physics and the fundamental forces that control subatomic matter.

Quantum potential. Mode of description of quantum systems favoured by Bohm, Hiley and co-workers in which the erratic and unpredictable fluctuations associated with quantum behaviour are regarded as a consequence of a 'potential' field analogous to, for example, the gravitational potential.

Relativity, theory of. The currently accepted description of space, time and motion, and a cornerstone of twentieth-century physics. The 'special' theory, first published by Einstein in 1905, introduced some unusual ideas such as time dilation and the equivalence of mass (m) and energy ($E = mc^2$). A key result of the special theory is that no material body, physical influence, or signal can exceed the speed of light (c). The later (1915) 'general' theory included the effects of gravitation on spacetime structure.

Schrödinger's cat paradox. A paradox arising from a thought experiment in which a quantum process is used to put a cat into an apparent superposition of live and dead states. (See p. 28 for full description.)

Schrödinger's equation. After Erwin Schrödinger, this is an equation, similar to that for a conventional wave, which describes the behaviour of the quantum wave function.

State function. Abstract mathematical object that encodes all the physical information needed to give the most complete available physical description of a quantum system. In many cases the state function can be represented as a wave function obeying Schrödinger's equation.

Two-slit experiment. An experiment first performed by Thomas Young in which light falls on two nearby narrow slits in a screen and produces an interference pattern on an image screen, thereby demonstrating the wave nature of light. (See p. 7 for a full description.)

Virtual particles. The Heisenberg uncertainty principle permits particles to appear and disappear again spontaneously, having survived for only very short durations. These fleeting entities are called 'virtual' to distinguish them from the more familiar, long-lived, 'real' particles.

Wave function. A mathematical object that describes the state of a quantum system. In simple cases the behaviour of the wave function is described by Schrödinger's equation.

Wave function, collapse or reduction of. The process that occurs when a measurement is made of a quantum system, whereby the wave function abruptly and discontinuously alters its structure. The significance of this 'collapse' is contentious.

Wave packet. Sometimes the wave function of a quantum system is concentrated in a narrow region of space. This configuration, which implies that the particle being described is relatively localized, is called a wave packet.

Zero point energy. An irreducible quantity of energy which, according to quantum mechanics, always resides in a system that is confined in some way. Its existence can be regarded as a consequence of Heisenberg's uncertainty principle.

FURTHER READING

T. Bastin (ed.), *Quantum Theory and Beyond* (Cambridge University Press, Cambridge, 1971).

D. Bohm, *Wholeness and the Implicate Order* (Routledge & Kegan Paul, London 1980).

J. F. Clauser and A. Shimony, 'Bell's theorem: experimental tests and implications' in *Reports on Progress in Physics* **41**, 1881–1927 (1978).

B. d'Espagnat, *The Conceptual Foundations of Quantum Mechanics* (Benjamin, New York, 1971); *In Search of Reality* (Springer-Verlag, New York, 1983); 'Quantum theory and reality' in *Scientific American*, November 1979, 158–81.

B. S. DeWitt, 'Quantum mechanics and reality', in *Physics Today*, September 1970, 30–5.

B. S. DeWitt and N. Graham, *The Many-Worlds Interpretation of Quantum Mechanics* (Princeton University Press, Princeton, N.J., 1973).

W. Heisenberg, *Physics and Philosophy* (Harper & Row, New York, 1959).

M. Jammer, *The Philosophy of Quantum Mechanics* (John Wiley, New York, 1974).

N. D. Mermin, 'Is the moon there when nobody looks? Reality and the quantum theory', in *Physics Today*, April 1985, 38–47.

A. I. M. Rae, *Quantum Physics: Illusion or Reality* (Cambridge University Press, 1986).

G. Ryle, *The Concept of Mind* (Barnes & Noble, London, 1949).

J. von Neumann, *Mathematical Foundations of Quantum Mechanics* (Princeton University Press, Princeton, N.J., 1955).

J. A. Wheeler and W. H. Zurek, *Quantum Theory and Measurement* (Princeton University Press, Princeton, N.J., 1983).

E. P. Wigner, 'Remarks on the mind-body question', in *The Scientist Speculates – An Anthology of Partly-Baked Ideas*, ed. I. J. Good, 284–302 (Basic Books, New York, 1962).

INDEX

Bold numbers indicate whole chapters